Oxidative Stress, Inflammation and Angiog
in the Metabolic Syndrome

Oxidative Stress, Inflammation and Angiogenesis in the Metabolic Syndrome

edited by

Raquel Soares
University of Porto, Portugal

and

Carla Costa
University of Porto, Portugal

Springer

Editors

Dr. Raquel Soares
Department of Biochemistry (U38/CFT)
Faculty of Medicine
University of Porto
Al. Prof. Hernani Monteiro
4200-319 Postage
Portugal
raqsoa@med.up.pt

Dr. Carla Costa
Department of Biochemistry (U38/FCT)
Laboratory for Molecular Cell Biology
Faculty of Medicine
University of Porto
Al. Prof. Hernani Monteiro
4200-319 Postage
Portugal
carcosta@med.up.pt

ISBN 978-1-4020-9700-3 e-ISBN 978-1-4020-9701-0

DOI 10.1007/978-1-4020-9701-0

Library of Congress Control Number: 2008942777

Printed on acid-free paper

9 8 7 6 5 4 3 2 1

springer.com

Preface

The metabolic syndrome (MS), a cluster of several risk factors for diabetes and cardiovascular disease, major causes of morbidity and death, is a highly prevalent condition in the western world. Ethiopathogenesis of MS is highly complex, and participating factors multiple and highly varied. Despite availability of diverse treatment tools and methodologies, the problem persists in its increasing tendency. This fact shows the necessity to improve the knowledge on MS, both in its overall complexity and in its biochemical mechanisms. An increasing number of studies confirm that oxidative stress, chronic inflammation and angiogenesis all play important roles in the pathogenesis of the MS. Although many growth factors and cytokines have been reported to interfere somehow in those entities, the precise interplay of those effectors among them and towards the MS is not yet clear.

Most of the published studies focus on the epidemiology, clinical symptoms, association between features of the MS, or the respective prevention/treatment strategies. In the present book, the knowledge gathered on MS as a whole, as well as on the implication of mechanisms of oxidative stress, chronic inflammation and angiogenesis in its development and progression is critically reviewed and discussed. It, thus, allows an integrated view of the condition, favouring a holistic approach towards preventive and therapeutic possibilities.

A special focus is put on some specific issues of recent discovery and/or progress. These include the role of glucose transporters within MS; the described effects of polyphenols as anti-oxidant, anti-inflammatory and anti-angiogenic compounds, driving them towards putative therapeutic strategies; or the role of NFκB, nitric oxide synthases, hypoxia-inducible factors, and many other molecules playing a role in the development of oxidative stress, inflammation, as well as angiogenesis. Given the novel concepts on oxidative stress highlighted recently, a chapter presenting an update of this issue is also included.

This book is written in order to fill the gap between basic science and medical care, and provide the reader, particularly under- and post-graduate students of health sciences, with skills to apply rigorous basic science to clinical settings of MS-associated disorders. We hope readers find it clear and simple, and at the same time stimulant and helpful. A most happy end would be its contribution for new research approaches to understand and help society solve this growing public health threat.

Porto, Portugal Raquel Soares
Porto, Portugal Isabel Azevedo

Contents

Contributors

Cassiano Abreu-Lima Department of Cardiology, Faculty of Medicine, University of Porto, Al. Prof. Hernâni Monteiro, 4200-319 Porto, Portugal, cassal@netcabo.pt

Ana Azevedo Department of Hygiene and Epidemiology, Cardiovascular R&D Unit, (U51/94/FCT), Faculty of Medicine, University of Porto, Al. Prof. Hernâni Monteiro, 4200-319 Porto, Portugal, anazev@med.up.pt

Isabel Azevedo Department of Biochemistry (U38/FCT), Faculty of Medicine, University of Porto, Porto, Portugal, isabelaz@med.up.pt

Conceição Calhau Department of Biochemistry (U38/FCT), Faculty of Medicine, University of Porto, Porto, Portugal, ccalhau@med.up.pt

Carla Costa Department of Biochemistry (U38/FCT) and Laboratory for Molecular Cell Biology, Faculty of Medicine, University of Porto, Porto, Portugal, carcosta@med.up.pt

Ana Faria Department of Biochemistry (U38/FCT), Faculty of Medicine, University of Porto, Porto, Portugal; Chemistry Investigation Centre (CIQ), Department of Chemistry, University of Porto 4169-007 Porto, Portugal, anafaria@med.up.pt

Elisa Keating Department of Biochemistry (U38/FCT), Faculty of Medicine, University of Porto, Al. Prof. Hernâni Motneiro, 4200-319 Porto, Portugal, keating@med.up.pt

João Laranjinha Center for Neurosciences and Cell Biology and Faculty of Pharmacy, University of Coimbra, Health Sciences Campus, Azinhaga de Santa Comba, 3000-548 Coimbra, Portugal laranjin@ci.uc.pt

Fatima Martel Department of Biochemistry (U38/FCT), Faculty of Medicine, University of Porto, Porto, Portugal, fmartel@med.up.pt

Rosário Monteiro Department of Biochemistry (U38/FCT), Faculty of Medicine, University of Porto, Porto, Portugal, rosariom@med.up.pt

Rita Negrão Department of Biochemistry (U38/FCT), Faculty of Medicine, University of Porto, Al. Prof. Hernâni Monteiro, 4200-319 Porto, Portugal, ritabsn@med.up.pt

Laura Ribeiro Department of Biochemistry (U38/FCT), Faculty of Medicine, University of Porto, Porto, Portugal, lribeiro@med.up.pt

Ana C. Santos Department of Hygiene and Epidemiology, Cardiovascular R&D Unit, (U51/94/FCT), Faculty of Medicine, University of Porto, Porto, Portugal, acsantos@med.up.pt

Alejandro Santos Department of Biochemistry (U38/FCT), Faculty of Medicine, University of Porto, Porto, Portugal, Faculty of Nutricional Sciences, University of Porto, 4200-319 Porto, Portugal, alejandr@med.up.pt

Raquel Soares Department of Biochemistry (U38/FCT), Faculty of Medicine, University of Porto, Porto, Portugal, raqsoa@med.up.pt

Chapter 1
The Metabolic Syndrome

Ana Azevedo, Ana C. Santos, Laura Ribeiro and Isabel Azevedo

Contents

Abstract Cardiovascular risk factors have a tendency to co-aggregate. One of the most well studied of these co-aggregations is the overlap between insulin resistance, obesity, hypertension and dyslipidaemia, now labelled as metabolic syndrome.

Although metabolic syndrome is multifactorial, there is a growing belief that visceral obesity may play an important role in the development of the syndrome. More recently, adipocyte hypertrophy is also getting much attention.

Susceptibility to the metabolic syndrome encompasses genetic factors and environmental conditions during early life, including intra-uterine time. Environmental factors, namely related with lifestyles – oscillation between work and rest, sleeping time, quality and quantity of food and meal schedule, social organization, level of stress and physical activity, play a paramount role in its causation and progressive development of superimposing vicious cycles. The consequences of the syndrome are many, with relevance to cardiovascular disease and diabetes. The proportion of affected people in the present world, together with the diverse nature of

A. Azevedo (✉)
Department of Hygiene and Epidemiology, Cardiovascular R&D Unit,
(U51/94/FCT), Faculty of Medicine, University of Porto, Al. Prof. Hernâni Monteiro,
4200-319 Porto, Portugal
e-mail: anazev@med.up.pt

R. Soares, C. Costa (eds.), *Oxidative Stress, Inflammation and Angiogenesis in the Metabolic Syndrome,* DOI 10.1007/978-1-4020-9701-0_1,

associating/causative factors, warrants a highly committed multidisciplinary effort from the whole society to deal with the problem.

Keywords Cardiovascular disease · Diabetes · Lifestyle · Metabolic syndrome · Stress

1.1 Concept and Components of the Metabolic Syndrome

Cardiovascular diseases are the leading cause of death worldwide, accounting for half of all deaths in middle age and one-third of all deaths in old age in most developed countries (World Health Organization 2007).

Cardiovascular risk factors have a tendency to co-aggregate across individuals and societies. One of the most well studied of these co-aggregations, difficult to disentangle, is the overlap between insulin resistance, obesity, hypertension and dyslipidaemia (Ogden et al. 2007; Rana et al. 2007; Shoelson et al. 2007). It has been recognized for several decades that individuals with insulin resistance often have hypertension, obesity and/or dyslipidaemia, and this cluster is also related with type 2 diabetes (Wingard et al. 1983). This risk factor clustering and its association with insulin resistance led to the proposal of a unique pathophysiological entity (Stern and Haffner 1986) now labelled as metabolic syndrome.

The metabolic syndrome, recognized as a major cause of type 2 diabetes and cardiovascular diseases, has become one of the major public health challenges worldwide (Eckel et al. 2005; Caterson et al. 2004; Galassi et al. 2006).The concept of a clinical entity composed by a constellation of metabolic disturbances was first proposed more than eighty years ago by Kylin, who described a cluster of hypertension, hyperglycaemia and gout (Kylin 1923). In 1947 Vague suggested that a particular obesity phenotype, then called android or male-type obesity, was associated with the metabolic disturbancies seen in diabetes and cardiovascular diseases (Vague 1947). The concept of syndrome X was labelled by Reaven in (1988), but the term metabolic syndrome, now well-established, is currently considered a better description of the situation.

The diversity of features associated with this condition and the study of the problem by different associations led to a variety of definitions. All share the inclusion of the main features – glucose intolerance, obesity, hypertension and dyslipidaemia, but differ in details and criteria. The most widely recognized definitions of the metabolic syndrome are those from the World Health Organization (WHO) (World Health Organization 1999), the European Group of Insulin Resistance (EGIR) (Balkau and Charles 1999), the National Cholesterol Education Program, Adult Treatment Panel III (ATP III) (Expert Pannel on Detection Evaluation and Treatment of High Blood Cholesterol in Adults 2001), the International Diabetes Federation (IDF) (Alberti et al. 2006), and the American Heart Association and the National Heart, Lung and Blood Institute (AHA/NHLBI) (Grundy et al. 2005).

The WHO definition considers insulin resistance as the core of the metabolic syndrome, the main difficulty in its application being the requirement of the euglycaemic clamp for the measurement of insulin resistance. Also focused on insulin resistance, but replacing the euglycaemic clamp by fasting insulin levels, easier to measure, is the EGIR diagnostic system. Whereas all the metabolic syndrome components are given the same importance in ATP III definition, a consensus group of IDF recognized central obesity as the core and necessary feature of the metabolic syndrome. Central obesity (intraabdominal or visceral obesity) is, thus, considered as a determinant in the causal pathway of the syndrome, impacting on the other components of the syndrome and on cardiovascular disease. Results from the most recent research do indeed support such a central and causal role of visceral adipocytes/adipose tissue in this pathogeny. As a matter of fact, visceral obesity seems to be the major predictor of incident metabolic syndrome, even when compared to insulin resistance (Palanippan et al. 2004; Tong et al. 2007; Srinivasan et al. 2002).

Due, in part, to differences in definition criteria, published prevalence estimates of metabolic syndrome vary broadly around the world, ranging from 10% to 50% (Cameron et al. 2004). An association with female gender and increasing age is consistently observed (Zimmet et al. 1997). Moreover, the prevalence is increasing, particularly in young individuals (Palanippan et al. 2004).

1.2 Pathogenesis – the Role of Visceral Adiposity

The pathogenesis of the metabolic syndrome is multiple and still poorly understood. No single factor has yet been identified as an underlying causal factor (Grundy et al. 2005). There is a growing belief, however, that obesity, specially visceral obesity, may play an important role in the development of the syndrome. Visceral adiposity seems to be an independent predictor of insulin sensitivity (Cnop et al. 2002; Wagenknecht et al. 2003; Katsuki et al. 2003), impaired glucose tolerance (Hayashi et al. 2003), elevated blood pressure (Rattarasarn et al. 2003; Bacha et al. 2003), and dyslipidaemia (Katsuki et al. 2003; Nieves et al. 2003).

Visceral fat is a highly active tissue from the metabolic point of view. It is apparently more susceptible to lipolysis than subcutaneous adipose tissue (van Harmelen et al. 2002), and produces, apparently at a higher magnitude than subcutaneous adipose tissue, various adipokines such as tumor necrosis factor-alpha (van Harmelen et al. 2002; Bertin et al. 2000), plasminogen activator inhibitor 1 (Alessi et al. 1997), interleukin 6 and C-reactive protein (You et al. 2008). On the other hand, it is a feabler producer of adiponectin, an adipokine more strongly correlated with subcutaneous fat (Cnop et al. 2003).

Bahceci and collaborators (Bahceci et al. 2007) found a positive correlation between adipocyte size and tumor necrosis factor-alpha, interleukin-6 and high-sensitivity C-reactive protein. On the other hand, adiponectin was found to be negatively correlated with adipocyte size. Although the adipocytes measured in this study were from subcutaneous adipose tissue, and the study conclusion pointed to an association between subcutaneous adipocyte size and inflammation, visceral obesity is generally considered to be much more pro-inflammatory (Santos et al. 2005). Its association with a variety of pathologies is much stronger than that of peripheral adiposity.

It has been shown that hypertension, obesity, and diabetes mellitus are independently associated with an increased risk of heart failure, regardless of coronary heart disease (Wisniacki et al. 2005; Kenchaiah et al. 2004; Kenchaiah et al. 2002; Kannel and McGee 1979; Scheuermann-Freestone et al. 2004; Arnlov et al. 2001; Bertoni et al. 2004). Azevedo and collaborators (Azevedo et al. 2007) found an association between systolic and diastolic dysfunction and the degree of metabolic syndrome, with the frequency and/or the severity of systolic and diastolic dysfunction increasing with the number of features of the metabolic syndrome. Importantly, early asymptomatic stages of cardiac dysfunction increased progressively with the severity of the metabolic syndrome, independently of systolic blood pressure.

Metabolic syndrome prevalence is high and its incidence continues to increase all over the world, as more and more people adopt the modern Western lifestyle (Carnethon et al. 2004; Wilsgaard et al. 2007). The main negative features of this lifestyle include stress (long-term and continuous, psychological), positive energy balance (excessive energy intake and low physical activity), low quality food (both energy dense and poor in micronutrients), and disruption of chronobiology. Genetic factors contribute to a varying degree.

Regulatory systems in human physiology are multiple, extremely complex and intermingled. An acute disturbance in any of them evokes reactions that tend to re-establish equilibrium. But when the stimuli, even if of moderate magnitude, tend to be highly repetitive or chronic, change and allostasis in one system impact on the other(s), and vicious cycles are created and steadily reinforced. When full-blown metabolic syndrome becomes apparent (at least three of defining features), it is only the tip of an iceberg that is coming to surface. A long history and many physiopathological events at multiple levels have occurred by then. Treatment of these patients requires a multidisciplinary approach, high commitment of both patient and therapists, and time. Prevention will be the best medicine.

As referred above, visceral obesity seems to play a central, and primordial, role in the metabolic syndrome. However, not all patients with the syndrome present this feature. This seems somehow contradictory to that putative primordial role. Perhaps as important as the amount of accumulated fat in the abdominal cavity, or even more important, is the size of abdominal adipocytes. We have shown that big adipocytes are more prone to rupture (Monteiro et al. 2006), and cell rupture will obviously constitute a focus of inflammation. Cinti and collaborators (Cinti et al. 2005) had already shown that macrophage crowns, in adipose tissue, surround dead adipocytes.

Subcutaneous adipocytes can also become hypertrophic, and most probably do also rupture. However abdominal adipocytes, besides being supported by much less dense connective tissue, when compared to subcutaneous adipose tissue, are subject to sudden pressure variations associated with cough, physical exercises (Cobb et al. 2005) and also sleep apnoea (Monteiro et al. 2007a). Intraabdominal pressure is higher in obese patients (Lambert et al. 2005), aggravating the situation.

Preadipocytes of upper-body obese women exhibit reduced differentiation and are more prone to apoptosis than preadipocytes isolated from adipose tissue of lower-body obese or lean women (Tchoukalova et al. 2007). The factors involved in this preadipocyte behaviour deserve investigation (Monteiro et al. 2007b).

In an interesting model of diet-induced metabolic syndrome, the fructose-fed rat, the animals exhibit a series of metabolic syndrome features but do not weigh more than controls. However, their abdominal adipocytes are hypertrophic, a modification prevented by blockade of the renin-angiotensin system (Furuhashi et al. 2004). Functional relationships between adipocytes and the renin-angiotensin-aldosterone system (Roberge et al. 2007) or between adipocytes and adrenocortex/aldosterone (Ehrhart-Bornstein et al. 2003) do probably deserve more attention.

Another characteristic of abdominal adipose tissue is its high metabolic activity and dense vascularisation. This high vascularisation is most probably due to the action of angiogenic/proinflammatory factors, of which leptin constitutes an example (Hausman and Richardson 2004; Wang et al. 2008). On the other hand, increased intra-abdominal pressure, as well as pressure variations, may easily create periods/zones of hypoxia, contributing to the production of hypoxic inducible factors. Vascular endothelial growth factor (VEGF), leptin, adenosine, and substance P, among others, will impact on other features of the metabolic syndrome. Furthermore, hypoxia by itself inactivates the adiponectin promoter (Hosogai et al. 2007).

Considering the importance of abdominal obesity and abdominal adipocyte hypertrophy in the metabolic syndrome, let us analyse what may lead to that dysmorphism.

As stated above, the metabolic syndrome shows a steady increase all over the world, accompanying acquisition of so-called modern Western lifestyle. And all the main negative features of this lifestyle (excessive energy intake, low physical activity, low quality food, stress and disruption of chronobiology) may directly impact on abdominal adiposity, beyond reverberating multiple indirect effects.

1.3 Energetic Balance

Feeding depends on a multitude of factors, both related to individual characteristics and socio-economic-cultural factors (Capaldi 1996). Food abundance, wide availability of energy dense meals and snacks, disruption of eating discipline (no time to go home for lunch, nobody to prepare the meals at home, work pressure superimposing on life habits, accelerating subjective time, availability of fast-food and machine disposers) easily lead people to an on-demand eating regime. Man, like other living beings, is endowed with complex homeostatic appetite control mechanisms which should be able to preclude excessive ingestion. But a series of other factors overwhelms those homeostatic mechanisms: excessive stress, with disturbed cortisol levels, anxiety and/or depressive feelings which may be compensated for by ingesting fast-absorvable carbohydrates, individualistic and hedonic culture (Berthoud 2007; Zheng and Berthoud 2007). However, homeostatic mechanisms may likewise apply: effort to get lacking micronutrients in a fast built diet, a hypothesis not yet tested.

Information and education, including social dimensions, seem paramount to change inadequate eating behaviour. As a matter of fact, learning can be included in homeostatic mechanisms (Capaldi 1996).

Energetic balance implies confront between energy ingestion and energy expenditure. Although the type of food may also impact on energy expenditure, thermogenesis has an energy cost, and many factors influence metabolism. The highest modifiable energy consumer is physical activity (Laaksonen 2002; Rennie 2003; Simmons et al. 2008). The extension and level of obligatory physical activity decreased dramatically with modern life commodities, and spontaneous activity did also decrease in parallel with dumped chronobiologic oscillations, permanent fed state, and amount of time dedicated to non-active leisure programs (TV, Internet, electronic games). Furthermore, physical activity directly impacts on various metabolic events beyond simple arithmetic of energy balance, such as activation of neurohumoral systems, alteration of hormone secretion, insulin sensitivity and lipolysis.

Energy expenses with thermogenesis did also decrease in modern affluent societies with generalized use of temperature homogenizers.

1.4 The Role of Stress

Although high energy intake and low energy expenditure are usually considered the main elements responsible for the present obesity crisis, "there are more roads to explore" (Keith et al. 2006).

1.4.1 Hypothalamus-Pituitary-Adrenal Cortex and Corticoids

One of the most important seems to be the level of chronic psychological stress. Stress leads to the activation of the two important neurohumoral systems, the hypothalamus-pituitary-adrenal cortex (HPA) axis, which has cortisol as its main effector, and the sympatho-adrenomedullary axis, which involves the release of catecholamines, among others. The survival value of these systems has been paramount along evolution. However, under the effect of modern life strains upon these systems, mainly when imposed in a chronic way, they are no longer adaptive, leading instead to dysregulation, allostasis and disease. For a thoughtful consideration of these ill aspects of modern life, the work of James Henry, collaborators and followers, both with animal and human communities (Henry and Stephens 1977; Folkow et al. 1997; Henry and Grim 1990; Folkow 2006), is highly recommended. They show how the repeated defence (alarm) reaction, which acutely prepares for attack or flight, when not followed by the naturally subsequent intense physical activity, becomes related with arterial hypertension. On the other hand, the defeat-submission reaction, when there is loss of control and social support, leads to metabolic syndrome related metabolic abnormalities. Social determinants of health, or the effect of social organization on stress and health (Brunner and Marmot 2006) have been carefully analysed and should be taken into consideration in any effort to understand the metabolic syndrome (Bjorntorp 1991; Bosma et al. 1997; Brunner 1997; Brydon et al. 2004; Siegrist 1996; Steptoe and Marmot 2002).

Cushing disease or cushingoid syndromes, with their characteristic metabolic abnormalities and central obesity, constitute a good model for the consequences of an excessive production or activity of cortisol. There is indeed a strong relationship between the HPA system and metabolism (Nieuwenhuizen and Rutters. 2008). That relationship includes expression of the enzyme 11beta-hydroxysteroid dehydrogenase type-1, which can generate active cortisol from inactive cortisone, and which is increased in adipose tissue with age (Li et al. 2007), in obesity (Desbriere et al. 2006), and further increased after exposure to cortisol and insulin (Bujalska et al. 1997). Bujalska et al. showed that 11beta-hydroxysteroid dehydrogenase type-1 gene was one of the most upregulated genes in omental but not subcutaneous preadipocytes after exposure to cortisol (Bujalska et al. 2006). In this study, glucocorticoid receptor expression was shown to be similar among omental and subcutaneous adipocytes, and not regulated by glucocorticoids. Cortisol production by omental adipocytes acts locally, as splanchnic cortisol production is not altered by obesity (Basu et al. 2005). Most interesting has been the demonstration that omental 11beta-hydroxysteroid dehydrogenase type-1 correlates with fat cell size independently of obesity (Michailidou et al. 2007), addressing the strain to adipocyte size.

A very recent publication attributes the mediation of glucocorticoid metabolic effects to AMP-activated protein kinase, a crucially central enzyme in the control of metabolism (Christ-Crain et al. 2008). On the other hand, mineralocorticoid receptors, which also bind glucocorticoids, seem to be involved in both adipogenesis, obesity and other metabolic syndrome features (de Paula et al. 2004; Caprio et al. 2007; Guo et al. 2008). Furthermore, adipocytes secrete factors that stimulate adrenal mineralocorticoid release and sensitize the adrenal cortex to angiotensin II (Krug and Ehrhart-Bornstein 2008).

In conclusion, besides causing direct metabolic effects known for a long time in the context of the stress response, corticosteroids influence the expression of 11beta-hydroxysteroid dehydrogenase type-1 in omental adipocytes, more so when those adipocytes are hypertrophic. Metabolic alterations reactive to cortisol tend, thus, to be feed-forward and may end in an endless spiral.

1.4.2 The Sympatho-Adrenomedullary Axis and Catecholamines

The second classic stress axis is the sympatho-adrenal system. This system is likewise activated in emergency situations, it responds also to psychological stimuli, and plays an essential role in the regulation of metabolic and cardiovascular homeostasis. Since the identification of noradrenaline as the primary neurotransmitter released from sympathetic nerves (von Euler 1948), this amine has received the largest part of attention, in a system which has many more agents to be considered, namely adrenaline, ATP and several peptides. Although these two catecholamines, adrenaline and noradrenaline, have different actions at some sites, their actions are very similar at other sites, and it is rather common to see them addressed, in the literature, simply as catecholamines as if they were more or less the same. However, when we analyse the role of these amines in the metabolic syndrome, we see that they behave very differently: whereas production/levels of noradrenaline associate positively with obesity and cardiovascular risk, adrenaline shows an inverse association with cardiovascular mortality (Christensen and Schultz-Larsen 1994; Reims et al. 2005). A recent publication on a long-term prospective study, in Norway, shows that the adrenaline response to a mental stress test is a negative predictor of future body mass index (BMI), waist circumference and triceps skinfold thickness after 18 years of follow-up (Flaa et al. 2008).

Whereas noradrenaline release from sympathetic nerve terminals is evoked by central sympathetic outflow, adrenal medulla responds to that same central command according to a large range of local and historical factors. To begin with, the proportion of adrenaline and noradrenaline cells in adrenal medulla is highly variable, probably in relation with genetic and development factors. Local, paracrine agents such as cortisol, somatostatin and others, determine the level of expression of the PNMT (phenylethanolamine-N-methyl transferase), the enzyme which converts noradrenaline into adrenaline (Wurtman and Axelrod 1965), the release of adrenaline (Kyetnanský et al. 1993; Park et al. 2008), or the differential release of adrenaline and noradrenaline (Ribeiro et al. 2004).

Recently, much attention has been devoted to the impact of COMT (catechol-O-methyltransferase) polymorphisms on cardiovascular health. One of the COMT gene polymorphisms, val158met, exerts a considerable influence on enzymic activity (Lachman et al. 1996) and is related with cardiovascular risk (Voutilanen et al. 2007; Annerbrink et al. 2008). Although COMT is involved in the metabolism of various compounds, including estrogens and polyphenols, it plays a most important role in the metabolism of catecholamines, much more of adrenaline than of noradrenaline. Released noradrenaline is for the most part taken up into the nerve terminals, where metabolism occurs primarily through MAO (monoaminoxidase). As to adrenaline, mostly released from the adrenals into the circulation, it is taken up through extraneuronal uptake system (EMT) and metabolised by COMT (Trendelenburg 1980; Martel et al. 1999).

The inverse relation between cardiovascular risk and both adrenaline levels (more adrenaline – lower risk) and COMT gene polymorphism (high activity – lower risk) seemed paradoxical, and pointed to the possibility of further intervenients in that relationship. As the product of COMT action upon adrenaline is metanephrine, we are considering the hypothesis of metanephrine having cardiovascular protective effects. Curiously, the primordial work of Eisenhofer et al. (1995) showed that 91% of plasma metanephrine was directly produced and released from the adrenal gland. Moreover, metanephrine outflow of all inquired organs except adrenals (heart, forearm, lungs, kidneys, mesenteric organs and liver) was always of a lesser magnitude than metanephrine inflow, indicating a net removal of metanephrine by the organs, instead of a net production were it only a metabolic product. We thus propose that metanephrine should be considered an hormone, not only an adrenaline metabolite, and its physiological effects looked after. Not much is done yet, but an inhibitory effect on platelet aggregation has already been described, suggesting an alpha 2-adrenoceptor antagonist effect (Werle et al. 1988; Lenz et al. 1991). On the other hand, a previous observation (Sarmento et al. 1984) suggested that metanephrine may inhibit nucleoside transport, what may be of physiological relevance in the putative potentiation of adenosine effects.

Corticosteroids are classic inhibitors of extraneuronal uptake of catecholamines. We have demonstrated that pharmacological blockade of COMT activity in the presence of corticosterone leads to a marked accumulation of ^{3}H-isoprenaline taken up from the extracellular space in the sympathetic nerve terminals (Azevedo and Osswald 1976). We therefore hypothesize that under circumstances of a low-activity COMT polymorphism (genetic factor) and chronic stress with high cortisol levels (environmental factor), adrenaline may accumulate in the sympathetic nerve terminals. Upon nerve stimulation, this amine will be released and may reinforce neurotransmitter release through action upon presynaptic facillitatory beta$_2$ adrenoceptors (Guimarães and Moura 2001). This mechanism may amplify cardiovascular responses by stress.

Considering adrenaline protective effects upon metabolic and cardiovascular health, it is interesting to remember that stimuli known to favour adrenaline release – physical exercise, hypoxia and hypoglycaemia, are by themselves known to associate with protective lifestyles.

1.4.3 Substance P

Stress reactions are multiple, and so is the number of involved systems and mediators. We presented a brief review of the main two (hypothalamus-pituitary-adrenal/corticosteroids and sympathoadrenal/catecholamines). Among the agents released by the sympathoadrenal system there is one peptide, substance P, also associated with other nervous, humoral and cellular systems, whose putative importance we would like to underline.

Substance P is an 11-amino acid member of the tachykinin peptide family (Chang and Leeman 1970), expressed in the central nervous system as well as in peripheral tissues, including the gastrointestinal, respiratory, urinary and immune systems, blood vessels, skin and adrenal gland (Lai et al. 1998; Pascual and Bost 1990; Severini et al. 2002; Murabayashi et al. 2007). In 1998 Maier and Watkins published a seminal paper proposing that in parallel with the causation flow proceeding from psychological and behavioural events to changes in the brain, consequent alterations in autonomic and endocrine outflow from the brain, to cells of the immune system, there is a communication network that flows from the immune system to the brain

(Maier and Watkins 1998). They further proposed that fight-flight evolved later and coopted this immune-brain circuitry. Recently, Rosenkranz (Rosenkranz 2007) reviewed the evidence supporting a central role for substance P in that bidirectional communication, i.e., on the possibility that substance P dysregulation may be a point of convergence underlying the overlap of chronic inflammatory disease and mood and anxiety disorders. Interestingly, substance P was designated "chemical stressor", and proposed to play an important role in the transition between acute and chronic stress (Mello et al. 2007). It increases neurokinin 1 receptor (receptor of substance P) expression in human mesenteric preadipocytes, and through activation of these receptors it induces interleukin-8 expression through a NF-kB dependent pathway (Karagiannides et al. 2006). This same team has just described an anti-obesity effect for a substance P antagonist, apparently acting through a decrease in appetite (Karagiannides et al. 2008).

Substance P appears as an interesting candidate to investigate in metabolic syndrome development. Its role in brain-immune system communication suggests a relevant presence in women, known to be much more prone to both immunological system diseases and to depression than men. As already stated above, the metabolic syndrome is also more prevalent in women, and appears to develop in this gender through a somewhat different way from what occurs in men (Table 1.1).

Substance P increases in the plasma of patients with liver disease (Lee et al. 1997; El-Raziky et al. 2005), due to the importance of the liver in its metabolism. As substance P is a potent vasodilator, its increase evokes an increase in renin and aldosterone levels. The rebound activation of this vasoconstrictor system in liver patients constitutes the pathophysiological mechanism of arterial hypertension frequently developing in these patients (El-Raziky et al. 2005). On the other hand, the liver is frequently affected by the metabolic alterations of the syndrome, through the so-called non-alcoholic fatty liver disease. Elevations in markers of liver injury, namely alanine aminotransferase and aspartate aminotransferase, have been claimed to add predictive power to C-reactive protein levels in identifying patients at risk of developing diabetes (Hanley et al. 2004; Haffner 2006). The possibility of liver injury, at a certain point in the syndrome evolution, contributing to the reinforcement of the plasma levels of substance P is another hypothesis deserving investigation.

1.5 Chronobiology

The last marked alteration in lifestyle that modern life brought and we would like to address here is related to disturbance of circadian rhythms. Life on earth has been always governed by the succession of day and night. There is great variety, but each species has its own rhythm according to oscillation between light and dark. Recent research in this topic revealed that cyclic function in living beings involves every organ and system, attaining the most recondite molecular organization. In mammals the rhythms are generated by a central clock, in the suprachiasmatic nucleus of the hypothalamus, which synchronizes numerous peripheral clocks in many, probably all tissues (Dunlap 1999; Roenneberg and Merrow 2003; Ptitsyn et al. 2006). It

Table 1.1 Characteristics of the study sample and prevalence of the metabolic syndrome and its defining features

	Women n = 416	Men n = 268
Age (years), mean (standard deviation)	61 (10)	62 (11)
Metabolic syndrome, n(%)	97 (23.3)	38 (14.2)
Number of features of the metabolic syndrome, n(%)	59 (14.2)	35 (13.1)
0		
1	123 (29.6)	106 (39.6)
High blood pressure	92 (60.9)	11 (23.9)
High triglycerides	4 (6.3)	11 (23.9)
Low HDL-cholesterol	20 (25.3)	5 (12.5)
Waist circumference	5 (7.8)	0
High fasting serum glucose	2 (3.3)	3 (7.9)
2	137 (32.9)	89 (33.2)
3	55 (13.2)	24 (9.0)
4–5	42 (10.1)	14 (5.2)
Features of the metabolic syndrome, n (%)	307 (73.8)	212 (79.1)
High blood pressure (≥ 130/85 mmHg)		
High triglycerides (≥ 150 mg/dl)	90 (21.6)	79 (29.5)
Low HDL-cholesterol (< 50 mg/dl women; < 40 mg/dl men)	66 (15.9)	23 (8.6)
Waist circumference > 88 cm women or > 102 cm men	215 (51.7)	45 (16.8)
High fasting serum glucose (≥ 110 mg/dl)	60 (14.4)	55 (20.5)
Coronary heart disease, n(%)	34 (8.2)	30 (11.2)
10-year risk of CHD (FHS risk score)*	104 (30.7)	20 (9.7)
< 5%		
5–10%	145 (43.3)	58 (28.2)
10–15%	57 (17.0)	57 (27.7)
≥ 15%	30 (9.0)	71 (34.5)
LVSD, n(%)	10 (2.4)	20 (7.5)
LV diameter/height (mm/m), median (interquartile range)	30.1 (28.3–32.2)	29.1 (27.1–31.2)
LV mass/height (g/m), median (interquartile range)	103.7 (86.4–124.5)	112.5 (92.0–136.1)
LV mass/height^{27} (g/m$^{2.7}$), median (interquartile range)	50.2 (40.8–61.0)	46.6 (37.7–57.2)
LV mass/FFM (g/kg), median (interquartile range)	3.97 (3.28–4.71)	3.51 (2.87–4.20)
Posterior wall/height (mm/m), median (interquartile range)	5.5 (5.1–6.2)	5.4 (4.8–6.0)
Interventricular septum/height (mm/m), median (interquartile range)	6.2 (5.6–6.8)	5.9 (5.4–6.6)
Relative wall thickness, median (interquartile range)	0.37 (0.33–0.42)	0.37 (0.33–0.41)
Left atrium/height (mm/m), median (interquartile range)	23.0 (21.3–24.8)	21.9 (20.5–23.7)
E wave/A wave (peak velocity), median (interquartile range)	0.95 (0.78–1.20)	0.94 (0.77–1.16)
Diastolic dysfunction, n(%)	97 (24.1)	74 (29.0)
Stage C of heart failure (symptomatic cardiac dysfunction), n(%)	35 (8.4)	14 (5.2)

*Valid values available for 541 participants.

CHD, coronary heart disease; FHS, Framingham Heart Study; LVSD, left ventricular systolic dysfunction; LV, left ventricle; FFM, fat-free mass

From: Azevedo et al. 2007.

appears that at least 20% and possibly up to 100% of transcripts in the organism are under circadian control (Zvonic et al. 2006; Ptitsyn et al. 2007; Holzberg and Albrecht 2003).

Both central and peripheral clocks are autonomous, self-sustaining oscillators. However, their rhythm is not exactly of 24-hour periods but around (circa) 24 hours. They are entrained by external signals: the suprachiasmatic nucleus by light, the peripheral clocks by the central clock and also by feeding (Roenneberg and Merrow 2003; Dunlap. 2006). After due consideration to the present knowledge of chronobiology, it will be easy to understand the effects that attenuation of the day/night behavioural differences, facilitated by artificial light, and disruption of feeding schedules may have on metabolism and health. Sleep duration is being reduced and this has been associated with metabolic and endocrine dysfunction (Morgan et al. 1998; Spiegel et al. 2004; Spiegel et al. 2005). At high levels of stress (Foster 2005), the reparative function of sleep seems of high importance (Åkersted 2006). Shift work has increased, and it has been shown to impact on metabolism and cardiovascular risk (Karlsson et al. 2003; Knutsson 2003; Ha and Park 2005). An intimate relationship has been demonstrated between clock genes, metabolism, adipocytes and obesity (Calvani et al. 2004; Ando et al. 2005; Turek et al. 2005; Staels 2006; Yang et al. 2006; Bray and Young 2007; Prasai et al. 2008).

Society pressures are impending on individual health through various factors and mechanisms, from a very early age, including intra-uterine life (Reilly et al. 2005; Berne and Björntorp 2006). Analysis of obesities with a developmental origin led to various different hypotheses of causes and mechanisms, including alterations in vagal sensory development (Fox and Murphy 2008).

As sympathetic/parasympathetic predominance accompanies oscillations in time and activity, as well as oscillations in metabolism, this paired system physiology perturbation may also contribute to the metabolic syndrome pathogenesis (Balbo et al. 2002; Kreier et al. 2003; Perciaccante et al. 2006; Chen et al. 2008).

1.6 Concluding Remarks

In conclusion, the metabolic syndrome refers to an aggregation of metabolic features appearing in variable order, whose developmental story seems to be different between men and women, and which has visceral adiposity and insulin resistance at its core. Environmental factors, namely related with lifestyle – oscillation between work and rest, sleeping time, quality and quantity of food and meal schedule, social organization and level of stress, education, self-esteem and physical activity, have a paramount role in its causation and progressive development of superimposing vicious circles. Susceptibility to the metabolic syndrome encompasses genetic factors and environmental conditions during early life, including intra-uterine time. The consequences of the syndrome are many, with relevance to cardiovascular disease and diabetes. The proportion of affected people in the present world, together with the diverse nature of associating/causative factors warrant a highly committed multidisciplinary effort from the whole society to deal with the problem.

Acknowledgments Research work supported by FCT (POCI, Programa Comunitário de Apoio and FEDER).

References

Åkersted T. *Stress, sleep and restitution.* In: Bengt B Arnetz and Rolf Ekman (eds) Stress in health and disease, Wiley-VCH, 2006.

Alberti K, Zimmet P, Shaw J. *Metabolic syndrome – a new world-wide definition. A consensus Statement from the International Diabetes Federation.* Diabet Med. 2006; 23: 469–80.

Alessi M, Peiretti F, Morange P, Henry M, Nalbone G, Juhan-Vague I. *Production of plasminogen activator inhibitor 1 by human adipose tissue: possible link between visceral fat accumulation and vascular disease.* Diabetes. 1997; 46: 860–7.

Arnlov J, Lind L, Zethelius B, Andren B, Hales CN, Vessby B, Lithell H. *Several factors associated with the insulin resistance syndrome are predictors of left ventricular systolic dysfunction in a male population after 20 years of follow-up.* Am Heart J. 2001; 142: 720–4.

Ando H, Yanagihara H, Hayashi Y et al. *Rhythmic mRNA expression of clock genes and adipocytokines in mouse visceral adipose tissue.* Endocrinology. 2005; 146: 5631–6.

Annerbrink K, Westberg L, Nilsson S et al. *Catechol O-methyltransferase val158-met polymorphism is associated with abdominal obesity and blood pressure in men.* Metabolism. 2008; 57: 708–11.

Azevedo A, Bettencourt P, Almeida PB et al. *Increasing number of components of the metabolic syndrome and cardiac structural and functional abnormalities – cross-sectional study of the general population.* BMC Cardiovascular Disorders. 2007; 7: 17. doi:10.1186/1471-2261-7-17.

Azevedo I, Osswald W. *Uptake, distribution and metabolism of isoprenaline in the dog saphenous vein.* Naunyn-Schmiedeberg's Arch Pharmacol. 1976; 295: 141–7.

Bacha F, Saad R, Gungor N, Janosky J, Arslanian SA. *Obesity, regional fat distribution, and syndrome X in obese black versus white adolescents: race differential in diabetogenic and atherogenic risk factors.* J Clin Endocrinol Metab. 2003; 88: 2534–40.

Bahceci M, Gokalp D, Bahceci S, Tuzcu A, Atmaca S, Arikan S. *The correlation between adiposity and adiponectin, tumor necrosis factor alpha, interleukin-6 and high sensitivity C-reactive protein levels. Is adipocyte size associated with inflammation in adults?* J Endocrinol Invest. 2007; 30: 210–4.

Balbo SL, Bonfleur ML, Carneiro EM et al. *Parasympathetic activity changes insulin response to glucose and neurotransmitters.* Diabetes Metab. 2002; 28: 3S13–7.

Balkau B, Charles MA. *Comment on the provisional report from the WHO consultation. European Group for the Study of Insulin Resistance (EGIR).* Diabet Med. 1999; 16: 442–3.

Basu R, Singh RJ, Basu A et al. *Obesity and type 2 diabetes do not alter splanchnic cortisol production in humans.* J Clin Endocrinol Metab. 2005; 90: 3919–26.

Berne C, Björntorp P. *The metabolic syndrome.* In: Bengt B Arnetz and Rolf Ekman (eds) Stress in health and disease, Wiley-VCH, 2006.

Berthoud H-R. *Interactions between the "cognitive" and "metabolic" brain in the control of food intake.* Physiol Behav. 2007; 91: 486–98.

Bertin E, Nguyen P, Guenounou M, Durlach V, Potron G, Leutenegger M. *Plasma levels of tumor necrosis factor-alpha (TNF-alpha) are essentially dependent on visceral fat amount in type 2 diabetic patients.* Diabetes Metab. 2000; 26: 178–82.

Bertoni AG, Hundley WG, Massing MW, Bonds DE, Burke GL, Goff DC Jr. *Heart failure prevalence, incidence, and mortality in the elderly with diabetes.* Diabetes Care. 2004; 27: 699–703.

Bjorntorp P. *Visceral fat accumulation: the missing link between psychosocial factors and cardiovascular disease?* J Int Med. 1991; 230: 195–201.

Bosma H, Marmot MG, Hemingway H et al. *Low job control and risk of coronary heart disease in the Whitehall II (prospective cohort) study.* BMJ. 1997; 314: 558–65.

Bray MS, Young ME. *Circadian rhythms in the development of obesity: potential role for the circadian clock within the adipocyte*. Obes Rev. 2007; 8: 169–81.
Brunner EJ. *Stress and the biology of inequality*. BMJ. 1997; 314: 1472.
Brunner E, Marmot M. *Social organization, stress, and health*. In: Michael Marmot, Richard G Wilkinson (eds) Social Determinants of health, 2nd ed., Oxford University Press, 2006; pp. 6–30.
Brydon L, Edwards S, Mohamed-Ali V, Steptoe A. *Socioeconomic status and stress-induced increases in interleukin-6*. Brain Behav Immun. 2004; 18: 281–90.
Bujalska IJ, Kumar S, Stewart PM. *Does central obesity reflect "Cushing's disease of the omentum"?* Lancet. 1997; 349: 1210–3.
Bujalska IJ, Quinkler M, Tomlinson JW et al. *Expression profiling of 11β-hydroxysteroid dehydrogenase type-1 and glucocorticoid-target genes in subcutaneous and omental human preadipocytes*. J Molec Endocrinol. 2006; 37: 327–40.
Calvani M, Scarfone A, Granato L et al. *Restoration of adiponectin pulsatility in severy obese subjects after weight loss*. Diabetes. 2004; 53: 939–47.
Cameron AJ, Shaw JE, Zimmet PZ. *The metabolic syndrome: prevalence in worldwide populations*. Endocrinol Metab Clin North Am. 2004; 33: 351–75.
Capaldi E. *Why we eat what we eat – The psychology of eating*. American Psychological Association, 1996.
Caprio M, Feve B, Claes A et al. *Pivotal role of the mineralocorticoid receptor in corticosteroid-induced adipogenesis*. FASEB J. 2007; 21: 2185–94.
Carnethon MR, Loria CM, Hill JO, Sidney S, Savage PJ, Liu K. *Risk factors for the metabolic syndrome: The coronary artery risk development in young adults (CARDIA) study, 1985-2001* Diabetes Care. 2004; 27: 2707–15.
Caterson ID, Hubbard V, Bray GA, Grunstein R, Hansen BC, Hong Y, et al. *Prevention Conference VII: Obesity, a worldwide Epidemic related to heart disease and stroke: Group III: Worldwide Comorbidities of Obesity*. Circulation. 2004; 110: 476–83.
Chang MM, Leeman SE. *Isolation of a sialogogic peptide from bovine hypothalamic tissue and its characterization as substance P*. J Biol Chem. 1970; 245: 4784–90.
Chen GY, Hsiao TJ, Lo HM, Kuo CD. *Abdominal obesity is associated with autonomic nervous derangement in healthy Asian obese subjects*. Clin Nutr. 2008; 27: 212–7.
Christ-Crain M, Kola B, Lolli F et al. *AMP-activated protein kinase mediates glucocorticoid-induced metabolic changes: a novel mechanism in Cushing's syndrome*. FASEB J. 2008; 22: 1672–83.
Christensen NJ, Schultz-Larsen K. *Resting venous plasma adrenalin in 70-year-old men correlated positively to survival in a population study: the significance of the physical working capacity*. J Intern Med. 1994; 235: 229–32.
Cinti S, Mitchell G, Barbatelli G et al. *Adipocyte death defines macrophage localization and function in adipose tissue of obese mice and humans*. J Lip Res. 2005; 46: 2347–55.
Cnop M, Havel PJ, Utzschneider KM, Carr DB, Sinha MK, Boyko EJ, et al. *Relationship of adiponectin to body fat distribution, insulin sensitivity and plasma lipoproteins: evidence for independent roles of age and sex*. Diabetologia. 2003; 46: 459–69.
Cnop M, Landchild MJ, Vidal J, Havel PJ, Knowles NG, Carr DR, et al. *The concurrent accumulation of intra-abdominal and subcutaneous fat explains the association between insulin resistence and plasma leptin concentrations. Distinct metabolic effects of two fat compartments*. Diabetes. 2002; 51: 1005–15.
Cobb WS, Burns JM, Kercher KW et al. *Normal intraabdominal pressure in healthy adults*. J Surg Res. 2005; 129: 231–5.
de Paula RB, da Silva AA, Hall JE. *Aldosterone antagonism attenuates obesity-induced hypertension and glomerular hyperfiltration*. Hypertension. 2004, 43: 41–7.
Desbriere R, Vuaroqueaux V, Achard V et al. *11beta-hydroxysteroid dehydrogenase type-1 mRNA is increased in both visceral and subcutaneous adipose tissue of obese patients*. Obesity (Silver Spring). 2006; 14: 794–8.

Dunlap JC. *Molecular bases for circadian clocks.* Cell. 1999; 96: 271–90.

Dunlap JC. *Physiology. Running a clock requires quality time together.* Science. 2006; 311: 184–6.

Eckel RH, Grundi SM, Zimmet PZ. *The metabolic syndrome.* Lancet. 2005; 365: 1415–28.

El-Raziky MS, Gohar N, El-Raziky M. *Substance P, rennin and aldosterone in chronic liver disease in Egyption children.* J Trop Pediatr 2005; 51: 320–3.

Ehrhart-Bornstein M, Lamounier-Zepter V, Schraven A et al. *Human adipocytes secrete mineralocorticoid-releasing factors.* Proc Natl Acad Sci. 2003; 100: 14211–6.

Eisenhofer G, Rundquist B, Åneman A et al. *Regional release and removal of catecholamines and extraneuronal metabolism to metanephrines.* J Clin Endocrinol Metab. 1995; 80: 3009–17.

Expert Pannel on Detection Evaluation and Treatment of High Blood Cholesterol in Adults. *Executive summary of the third report of the National Cholesterol Education Program (NCEP) expert panel on detection, evaluation, and treatment of high blood cholesterol in adults (Adult Treatment Panel III).* JAMA. 2001; 285: 2486–97.

Flaa A, Sandvik L, Kjeldsen SE et al. *Does sympathoadrenal activity predict changes in body fat? An 18-y follow-up study.* Am J Clin Nutr. 2008; 87: 1596–601.

Folkow B. *Evolutionary aspects of stress.* In: Bengt B Arnetz and Rolf Ekman (eds) Stress in health and disease, Wiley-VCH, 2006.

Folkow B, Schmidt T, Uvnäs-Moberg K (eds). *Stress, health and the social environment. James P Henry's ethological approach to medicine, reflected by recent research in animals and man.* Acta Physiol Scand 1997; 61 Suppl. 640: 1–179.

Foster RG, Wulff K. *Science and Society: The rhythm of rest and excess.* Nat Rev Neuroscience. 2005; 6: 407.

Fox EA, Murphy MC. *Factors regulating vagal sensory development: potential role in obesities of developmental origin.* Physiol Behav. 2008; 94: 90–104.

Furuhashi M, Ura N, Takizawa H et al. *Blockade of the renin-angiotensin system decreases adipocyte size with improvement in insulin sensitivity.* J Hypertens 2004; 22: 1977–82.

Galassi A, Reynolds K, He J. *Metabolic syndrome and risk of cardiovascular disease: a meta-analysis.* Am J Med. 2006; 119: 812–9.

Grundy SM, Cleeman JI, Daniels SR, Donato KA, Eckel RH, Franklin BA, et al. *Diagnosis and management of the metabolic syndrome: An American heart association/national heart, Lung, and Blood Institute Scientific Statement.* Circulation. 2005; 112: 2735–52.

Guo C, Ricchiuti V, Lian BQ, Yao TM, Coutinho P, Romero JR, Li J, Williams GH, Adler GK. *Mineralocorticoid receptor blockade reverses obesity-related changes in expression of adiponectin, peroxisome proliferator-activated receptor-gamma, and proinflammatory adipokines.* Circulation. 2008; 117:2253–61.

Guimarães S, Moura D. *Vascular adrenoceptors: an update.* Pharmacol Rev. 2001; 53: 319–56.

Ha M, Park J. *Shiftwork and metabolic risk factors of cardiovascular disease.* J Occup Health. 2005; 47: 89–95.

Haffner SM. *Relationship of metabolic risk factors and development of cardiovascular disease and diabetes.* Obesity. 2006; 14 Suppl June: 121S–127S.

Hanley AJ, Williams K, Festa A et al. *Elevations in markers of liver injury and risk of type 2 diabetes: the Insulin Resistance Atherosclerosis Study.* Diabetes. 2004; 53: 2623–32.

Hausman GJ, Richardson RL. *Adipose tissue angiogenesis.* J Anim Sci. 2004; 82: 925–34.

Hayashi T, Boyko EJ, Leonetti DL, McNeely MJ, Newell-Morris L, Kahn SE, et al. *Visceral adiposity and the risk of impaired glucose tolerance. A prospective study among Japanese Americans.* Diabetes Care. 2003; 26: 650–5.

Henry JP, Grim C. *Psychosocial mechanisms of primary hypertension.* Editorial. Hypertension. 1990; 8: 783–93.

Henry JP, Stephens PM (1977) *Stress, health and the psychosocial environment. A sociobiological approach to medicine.* New York, Springer Verlag.

Holzberg D, Albrecht U. *The circadian clock: a manager of biochemical processes within the organism.* J Neuroendocrinol. 2003; 15: 339–43.

Hosogai N, Fukuhara A, Oshima K et al. *Adipose tissue hypoxia in obesity and its impact on adipokine dysregulation.* Diabetes. 2007; 56: 901–11.

Kannel WB, McGee DL. *Diabetes and cardiovascular disease. The Framingham study.* JAMA. 1979; 241:2035–8.

Karagiannides I, Kokkotou E, Tansky M et al. *Induction of colitis causes inflammatory responses in fat depots: evidence for substance P pathways in human mesenteric preadipocytes.* Proc Natl Acad Sci. 2006; 103: 5207–12.

Karagiannides I, Torres D, Tseng Y-H et al. *Substance P as a novel anti-obesity target.* Gastroenterology. 2008; 134: 747–55.

Karlsson BH, Knutsson AK, Lindahl BO, Alfredsson LS. *Metabolic disturbances in male workers with rotating three-shift work: Results of the WOLF study.* Int Arch Occup Environ Health. 2003; 76: 424–30.

Katsuki A, Sumida Y, Urakawa H, Gabazza EC, Murashima S, Maruyama N, et al. *Increased visceral fat and serum levels of triglycerides are associated with insulin resistance in Japanese metabolically obese, normal weight subjects with normal glucose tolerance.* Diabetes Care. 2003; 26: 2341–4.

Keith SW, Redden DT, Katzmarzyc PT et al. *Putative contributers to the secular increase in obesity: exploring the roads less travelled.* Int J Obes. 2006; 30: 1585–94.

Kenchaiah S, Evans JC, Levy D, Wilson PWF, Benjamin EJ, Larson MG, Kannel WB, Vasan RS. *Obesity and the Risk of Heart Failure.* N Engl J Med. 2002; 347:305–13.

Kenchaiah S, Gaziano JM, Vasan RS. *Impact of obesity on the risk of heart failure and survival after the onset of heart failure.* Med Clin North Am. 2004; 88:1273–94.

Knutson A. *Health disorders of shift workers.* Occup Med (Lond). 2003; 53: 103.

Kreier F, Yilmaz A, Kalsbeek A et al. *Hypothesis: shifting the equilibrium from activity to food leads to autonomic unbalance and the metabolic syndrome.* Diabetes, 2003; 52: 2652–6.

Krug AW, Ehrhart-Bornstein M. *Adrenocortical dysfunction in obesity and the metabolic syndrome.* Horm Metab Res. 2008; 40: 515–7.

Kyetnanský R, Fukuhara K, Pacák K et al. *Endogenous glucocorticoids restrain catecholamine synthesis and release at rest and during immobilization stress in rats.* Endocrinology. 1993; 133: 1411–9.

Kylin E: *Studien über das Hypertonie-Hyperglyka "mie-Hyperurika" miesyndrom.* Zentralblatt für Innere Medizin. 1923; 44: 125.

Lachman HM, Papolos DF, Saito T et al. *Human catechol-O-methyltransferase pharmacogenetics: description of a functional polymorphism and its potential application to neuropsyquiatric disorders.* Pharmacogenetics. 1996; 6: 243–50.

Lambert DM, Marceau S, Forse RA. *Intra-abdominal pressure in the morbidly obese.* Obes Surg. 2005; 15: 1225–32.

Li X, Lindquist S, Chen R et al. *Depot-specific messenger RNA expression of 11beta-hydroxysteroid dehydrogenase type-1 and leptin in adipose tissue of children and adults.* Int J Obes (Lond). 2007; 31: 820–8.

Laaksonen DE, Lakka HM, Salonen JT, Niskanen LK, Rauramaa R, Lakka TA. *Low levels of leisure-time physical activity and cardiorespiratory fitness predict development of the metabolic syndrome.* Diabetes Care. 2002; 25:1612–8.

Lai JP, Douglas SD, Ho WZ. *Human lymphocytes express substance P and its receptor.* J Neuroimmunol. 1998; 86: 80–6.

Lee FY, Lin HC, Tsai YT et al. *Plasma substance P levels in patients with liver cirrhosis: relationship to systemic and portal hemodynamics.* Am J Gastroenterol. 1997; 92: 2080–4.

Lenz T, Werle E, Strobel G, Weicker H. *O-methylated and sulfoconjugated catecholamines: differential activities at human platelet alpha 2-adrenoceptors.* Can J Physiol Pharmacol. 1991; 69: 929–37.

Maier SF, Watkins LR. *Cytokines for psychologists: implications of bidirectional immune-to-brain communication for understanding behaviour, mood, and cognition.* Psychological Rev. 1998; 105: 83–107.

Martel F, Ribeiro L, Calhau C, Azevedo I. *Comparison between uptake$_2$ and rOCT1: effects of catecholamines, metanephrines and corticosterone*. Naunyn-Schmiedeberg's Arch Pharmacol. 1999; 359: 303–9.

Mello DM, Marcinichen DR, Madruga D et al. *Involvement of NK1 receptors in metabolic stress markers after the central administration of substance P.* Behav Brain Res. 2007; 181: 232–8.

Michailidou Z, Jensen MD, Dumesic DA et al. *Omental 11beta-hydroxysteroid dehydrogenase 1 correlates with fat cell size independently of obesity*. Obesity (Silver Spring). 2007; 15: 1155–63.

Monteiro R, Calhau C, Azevedo I. *Obstructive sleep apnoea and adipocyte death*. Eur J Heart Failure. 2007a; 9: 103–4.

Monteiro R, Calhau C, Azevedo I. *Comment on Tchoukalova Y, Koutsari C, Jensen M (2007b) Committed subcutaneous preadipocytes are reduced in human obesity*. Diabetologia 50: 151–157. Diabetologia. 2007; 50: 1569.

Monteiro R, de Castro PMST, Calhau C, Azevedo I. *Adipocyte size and liability to cell death*. Obes Surg. 2006; 16:804–6.

Morgan L, Arendt J, Owens D et al. *Effects of the endogeneous clock and sleep time on melatonin, insulin, glucose and lipid metabolism.* J Endocrinol. 1998; 157: 443–51.

Murabayashi H, Kuramoto H, Kawano H et al. *Immunohistochemical features of substance P-immunoreactive chromaffin cells and nerve fibers in the rat adrenal gland.* Arch Histol Cytol. 2007; 70: 183–96.

Nieuwenhuizen AG, Rutters F. *The hypothalamic-pituitary-adrenal axis in the regulation of energy balance*. Physiol Behav. 2008; 94: 169–74.

Nieves DJ, Cnop M, Retzlaff B, Walden CE, brunzell JD, Knopp RH, et al. *The atherogenic lipoprotein profile associated with obesity and insulin resistance is largely attributable to intra-abdominal fat*. Diabetes. 2003; 52: 172–9.

Ogden CL, Yanovski SZ, Carol MD, Flegal KM. *The epidemiology of obesity*. Gastroenterology. 2007; 132: 2087–102.

Palanippan L, Carnethon M, Wang Y, Hanley A, Fortmann S, Haffner S, et al. *Predictors of the incident metabolic syndrome in adults: the Insulin Resistance Atherosclerosis Study.* Diabetes Care. 2004; 27: 788–93.

Park YS, Choi YH, Park CH et al. *Non-genomic glucocorticoid effects on activity-dependent potentiation of catecholamine release in chromaffin cells*. Endocrinology. 2008; 149: 4921–7.

Pascual DW, Bost KL. *Substance P production by P388D1 macrophages: a possible autocrine function for this neuropeptide*. Immunology. 1990; 71: 52–6.

Perciaccante A, Fiorentini A, Paris A et al. *Circadian rhythm of the autonomic nervous system in insulin resistant subjects with normoglycemia, impaired fasting glycemia, impaired glucose tolerance, type 2 diabetes mellitus*. BMC Cardiovasc Disord. 2006; 6: 19.

Prasai MJ, George JT, Scott EM. *Molecular clocks, type 2 diabetes and cardiovascular disease.* Diabetes Vasc Dis Res. 2008; 5: 89–95.

Ptitsyn AA, Zvonic S, Conrad SA et al. *Circadian clocks resounding in peripheral tissues*. PLoS Comput Biol. 2006; 2: e16.

Ptitsyn AA, Zvonic S, Gimble JM. *Digital signal processing reveals circadian baseline oscillation in majority of mammalian genes*. PLoS Comput Biol. 2007; 3:e120.

Rana J, Nieuwdorp M, Jukema J, Kastelein J. *Cardiovascular metabolic syndrome – an interplay of obesity, inflammation, diabetes and coronary heart disease*. Diabetes Obes Metab. 2007; 9: 218–32.

Rattarasarn C, Leelawattana R, Soonthornpun S, Setasuban W, Thamprasit A, Lim A, et al. *Regional abdominal fat distribution in lean and obese thai type 2 diabetic women: relationships with insulin sensitivity amd cardiovascular risk factors*. Metabolism. 2003; 52: 1444–7.

Reaven GM. Banting lecture 1988. *Role of insulin resistance in human disease*. Diabetes. 1988; 37: 1595–607.

Reilly JJ, Armstrong J, Dorosty AR et al. *Early life risk factors for obesity in childhood: cohort study*. BMJ. 2005; 230: 1357.

Reims HM, Sevre K, Fossum E et al. *Adrenaline during mental stress in relation to fitness, metabolic risk factors and cardiovascular responses in young men.* Blood Press. 2005; 14: 217.

Rennie KL, McCarthy N, Yazdgerdi S, Marmot M, Brunner E. *Association of the metabolic syndrome with both vigorous and moderate physical activity.* Int J Epidemiol. 2003; 32: 600–6.

Ribeiro L, Martel F, Azevedo I *Effect of somatostatin on the release of adrenaline and noradrenaline from bovine adrenal chromaffin cells.* In: R Borges, L Gandia (Eds) Cell Biology of the chromaffin cell. University of La Laguna, Spain, 2004; pp. 1–4.

Roberge C, Carpentier AC, Langlois M-F et al. *Adrenocortical dysregulation as a major player in insulin resistance and onset of obesity.* Am J Physiol Endocrinol Metab. 2007; 293: E1465.

Roenneberg T, Merrow M. *The network of time: understanding the molecular circadian system.* Curr Biol. 2003; 13: R198.

Rosenkranz MA. *Substance P at the nexus of mind and body in chronic inflammation and affective disorders.* Psychological Bull. 2007; 6: 1007–37.

Santos A-C, Lopes C, Guimarães JT, Barros H. *Central obesity as a major determinant of increased high-sensitivity C-reactive protein in metabolic syndrome.* Intern J Obesity. 2005; 29: 1452–6.

Sarmento A, Albino Teixeira A, Azevedo I. *Decrease of 3H-uridine incorporation in the dog saphenous smooth muscle by isoprenaline or noradrenaline.* Arq Inst Farm Terap Exp Coimbra. 1984; 22: 145–56.

Scheuermann-Freestone M, Neubauer S, Clarke K. *Abnormal cardiac muscle function in heart failure is related to insulin resistance.* Cardiovasc J S Afr. 2004; 15(4 Suppl 1): S12.

Severini C, Improta G, Falconieri-Erspamer G et al. *The tachykinin peptide family.* Pharmacol Rev. 2002; 54: 285–322.

Shoelson SE, Herrero L, Naaz A. *Obesity, inflammation and insulin resistance.* Gastroenterology. 2007; 132: 2169–80.

Siegrist J. *Adverse health effects of high-effort/low-reward conditions.* J Occup Health Psychol. 1996; 1: 27–41.

Simmons RK, Griffin SJ, Steele R, Wareham NJ, Ekelund U, ProActive Research Team. *Increasing overall physical activity and aerobic fitness is associated with improvements in metabolic risk: cohort analysis of the ProActive trial.* Diabetologia. 2008; 51:787–94.

Spiegel K, Knutson K, Leproult R et al. *Sleep loss: a novel risk factor for insulin resistance and type 2 diabetes.* J Appl Physiol. 2005; 99: 2008–19.

Spiegel K, Tasali E, Penev P et al. *Brief communication: sleep curtailment in healthy young men is associated with decreased leptin levels, elevated ghrelin levels and increased hunger and appetite.* Ann Intern Med. 2004; 141: 846–50.

Srinivasan S, Myers L, Berenson G. *Predictability of childhood adiposity and insulin for developing insulin resistance syndrome (syndrome X) in young adulthood: the Bogalusa Heart Study.* Diabetes. 2002; 51: 204–9.

Staels B. *When the clock stops ticking, metabolic syndrome explodes.* Nat Med. 2006; 12: 54–5.

Steptoe A, Marmot M. *The role of psychobiological pathways in socio-economic inequalities in cardiovascular disease risk.* Eur Heart J. 2002; 23: 13–25.

Stern MP, Haffner SM. *Body fat distribution and hyperinsulinemia as risk factors for diabetes and cardiovascular disease.* Arteriosclerosis. 1986; 6: 123–30.

Tchoukalova Y, Koutsari C, Jensen M. *Committed subcutaneous preadipocytes are reduced in human obesity.* Diabetologia. 2007; 50: 151–7.

Tong J, Boyko E, Utzschneider K, McNeely M, Hayashi T, Carr D, et al. *Intra-abdominal fat accumulation predicts the development of the metabolic syndrome in non-diabetic Japanese-Americans.* Diabetologia. 2007; 50: 1156–60.

Trendelenburg U. *A kinetic analysis of the extraneuronal uptake and metabolism of catecholamines.* Rev Physiol Biochem Pharmacol. 1980; 87: 33–115.

Turek FW, Joshu C, Kohsaka A et al. *Obesity and metabolic syndrome in circadian clock mutant mice.* Science. 2005; 308: 1043–5.

Vague J. *La différenciation sexuelle, facteur determinant des formes de l' obésité.* Presse Med. 1947; 30: 339.

van Harmelen V, Dicker A, Rydén M, Hauner H, Lönnqvist F, Näslund E, et al. *Increased lipolysis and decreased leptin production by human omental as compared with subcutaneous preadipocytes.* Diabetes. 2002; 51: 2029–36.

von Euler U. *Identification of the sympathomimetic ergone in adrenergic nerves of cattle (sympathin N) with laevo-nor-adrenaline.* Acta Physiol Scand. 1948; 16: 63.

Voutilanen S, Tuomainen TP, Korhonen M et al. *Functional COMT Val158Met polymorphism, risk of acute coronary events and serum homocysteine: the Kuopio ischaemic heart disease risk factor study.* PLoS ONE. 2007; 2: e181.

Wagenknecht LE, Langefeld CD, Scherzinger AL, Norris JM, Haffner SM, Saad MF, et al. *Insulin sensitivity, insulin secretion, and abdominal fat: The Insulin Resistance Atherosclerosis Study (IRAS) Family Study.* Diabetes. 2003; 52: 2490–6.

Wang B, Wood I, Trayhurn P. *Hypoxia induces leptin gene expression and secretion in human preadipocytes: differential effects of hypoxia on adipokine expression by preadipocytes.* J Endocrinol. 2008; 198: 127–34.

Werle E, Michel G, Lenz T et al. *Restricted alpha- and beta-adrenoceptor affinity of sulfoconjugated catecholamines in human mononuclear leukocytes, platelets, and fat cells and reduction of the postreceptor mechanisms.* Int J Sports Med. 1988; 9: S93.

Wilsgaard T, Jacobsen BK. *Lifestyle factors and incident metabolic syndrome. The Tromsø study 1979-2001.* Diabetes Res Clin Pract. 2007; 78: 217–24.

Wingard D, Barrett-Connor E, Criqui M, Suarez L. *Clustering of heart disease risk factors in diabetic compared to non-diabetic adults.* Am J Epidemiol. 1983; 117: 19–26.

Wisniacki N, Taylor W, Lye M, Wilding JPH. *Insulin resistance and inflammatory activation in older patients with systolic and diastolic heart failure.* Heart. 2005; 91: 32–7.

World Health Organization. *World health statistics.* 2007 (cited 2008 May); Available from http://www.who.int/whosis/whostat2007_1mortality.pdf

Wurtman R, Axelrod J. *Adrenaline synthesis: Control by the pituitary gland and adrenal glucocorticoids.* Science. 1965; 150: 1464–5.

Yang X, Downes M, Yu RT et al. *Nuclear receptor expression links the circadian clock to metabolism.* Cell. 2006; 126: 801–10.

You T, Nicklas BJ, Ding J, Penninx BW, Goodpaster BH, Bauer DC, Tylavsky FA, Harris TB, Kritchevsky SB. *The metabolic syndrome is associated with circulating adipokines in older adults across a wide range of adiposity.* J Gerontol A Biol Sci Med Sci. 2008; 63: 414–9.

Zheng H, Berthoud H-R. *Eating for pleasure or calories.* Curr Opin Pharmacol. 2007; 7: 607–12.

Zimmet PZ, McCarty DJ, de Courten MP. *The global epidemiology of non-insulin-dependent diabetes mellitus and the metabolic syndrome.* J Diabetes Complications. 1997; 11: 60–8.

Zvonic S, Ptitsyn AA, Conrad SA et al. *Characterization of peripheral circadian clocks in adipose tissue.* Diabetes. 2006; 55: 962–70.

Chapter 2
Oxidative Stress: From the 1980's to Recent Update

João Laranjinha

Contents

Abstract Free radicals-mediated oxidation of biomolecules and oxidative stress were implausible notions just a few decades ago. The discovery of superoxide dismutase in the late 1960's triggered intense research that ultimately led to the description of the production of free radicals and oxidants by mitochondria and by several metabolic pathways in mammalian cells. In view of the threats imposed by free radicals and oxidants, life in an aerobic environment required antioxidant strategies to prevent and repair potential oxidative damage to vital cell components. As a corollary of these discoveries Helmut Sies formulated the concept of *oxidative stress*, emphasizing the balance in the dynamic equilibrium between oxidants and antioxidants. Later, it was recognized that free radicals and oxidants are not only noxious cellular stressors but also play an essential role in cellular signalling and redox regulation of metabolic process. In particular, it has been appreciated the involvement of free radical and oxidants in discreet redox pathways, suggesting that specific mechanisms have evolved for free radicals and oxidants signalling. This and other observations, arguing against a global imbalance between oxidants and antioxidants, led to an update of the notion of oxidative stress in order to emphasize discreet and compartmentalized cellular redox circuits. The updated concept of oxidative stress may thus help project novel therapeutic approaches selectively directed to targets and disease conditions.

J. Laranjinha (✉)
Center for Neurosciences and Cell Biology and Faculty of Pharmacy, University of Coimbra, Health Sciences Campus, Azinhaga de Santa Comba, 3000-548 Coimbra, Portugal
e-mail: laranjin@ci.uc.pt

R. Soares, C. Costa (eds.), *Oxidative Stress, Inflammation and Angiogenesis in the Metabolic Syndrome,* DOI 10.1007/978-1-4020-9701-0_2,

Keywords Antioxidants · Free radicals · Nitrosative stress · Oxidative stress · Redox signalling

2.1 Free Radicals in Biology and Medicine Come of Age

Deleterious oxidative processes mediated by free radicals have been involved in aging and in a vast array of diseases, including cancer, cardiovascular diseases, inflammation and neurodegenerative diseases, and yet free radicals-mediated oxidation of biomolecules and oxidative stress were implausible notions just a few decades ago.

In the 1950s, Rebecca Gerschman linked the toxic effects of oxygen to free radical formation (Gerschman et al. 1954) and Denham Harman postulated that aging was associated with the cumulative cellular damage induced by free radicals (Harman 1956). These notions, apparently, did not attract much interest from biologists and biochemists, and remained unnoticed for several years.

The point of view radically changed in 1969 when McCord and Fridovich purified an enzyme from bovine erythrocytes that catalyzes the dismutation of superoxide radical ($O_2^{\bullet -}$) to molecular oxygen and hydrogen peroxide (H_2O_2). The enzyme was named superoxide dismutase (SOD) (McCord and Fridovich 1969). Superoxide radical had been discovered in the 1930s by Linus Pauling (Pauling 1979) but was of concern only to chemists and, in the 1960s, free radicals were considered too short lived and reactive to play a role in life processes. The landmark discovery of SOD was prominent to bridge free radical chemistry to Biology and Medicine.

Superoxide radical was shown to be enzymatically produced by xanthine oxidase (Knowles et al. 1969) and a role in biological defense mechanisms was soon proposed (Babior et al. 1973). More generally, it was recognized that, in addition to the concerted tetravalent reduction of oxygen to water by cytochrome c oxidase in mitochondrial electron transport chain, side reactions encompassing the production of partial reduction products of molecular oxygen, some of free radical nature, was an inherent consequence to aerobic metabolism. Thus, a family of reactive intermediates resulting from the incomplete reduction of oxygen included superoxide radical (one electron) and hydrogen peroxide (2 electrons). Hydrogen peroxide does not contain unpaired electrons in the valence orbitals and, therefore, is not a free radical molecule but, upon lysis of the O—O bond by ferrous iron (known as Fenton reaction), yields the most powerful oxidant known in a biological setting, the hydroxyl radical, $HO^{\bullet}$. Conversely to superoxide radical and hydrogen peroxide, which are less reactive and, therefore, more selective in its targets, the hydroxyl radical, upon formation, oxidizes indiscriminately and site-specifically any biomolecule.

Collectively, free radicals and derived oxidants were coined as "reactive oxygen species (ROS)" in order to include free radicals (superoxide and hydroxyl radicals) as well as non-radicals, such as hydrogen peroxide and the electronically excited state of oxygen (singlet oxygen), which may be produced by photosensitization (e.g. in the presence of porphirins). In fact, it must be clearly seen that, in spite of the low reactivity, the triplet ground state of oxygen qualifies as free radical for it contains two unpaired electrons with parallel spin in distinct π-antibonding orbitals but a spin restriction creates a barrier to the insertion of a pair of electrons simultaneously, preventing its reaction with biomolecules.

The previously mentioned notions were supported by pioneer experiments showing that the mitochondrial respiratory chain can produce hydrogen peroxide (Loschen et al. 1971; Boveris et al. 1972; Loschen et al. 1973; Boveris and Chance 1973) and superoxide radical (Loschen et al. 1974; Boveris and Cadenas 1975). The existence of a superoxide dismutase located in the mitochondrial matrix (Weisiger and Fridovich 1973) indicated that superoxide radical could be efficiently dismutated to hydrogen peroxide, supporting its mitochondrial production. Moreover, it was demonstrated that the univalent reduction of oxygen that takes place in mammalian organs produces superoxide radicals at a rate of about 2% of the total oxygen uptake (Boveris et al. 1972).

In addition to the formation of ROS in the inner mitochondrial membrane, the outer membrane, and several metabolic pathways in mammalian cells, such as the microsomal electron transport can generate ROS (Fridovich 1978; Naqui et al. 1986). In particular, oxidases, such as the monoamine oxidase located at the outer mitochondrial membrane and amino acid oxidases of peroxisomes generate hydrogen peroxide during its catalytic cycle, as well as in microsomal electron transport chain, uncoupled nitric oxide synthase and cyclooxygenase (Naqui et al. 1986; Pryor 1986; Cross et al. 1987). Others, notably NADPH oxidase and xanthine dehydrogenase synthesize superoxide radical. The former participates in the mechanisms of viruses and bacteria neutralization by the immune system (via the respiratory burst) whereas the latter activity, via conversion to xanthine oxidase by thiol oxidation or irreversible proteolytic cleavage, was shown to increase the vascular production of superoxide radical in ischemia-reperfusion (Zweier et al. 1994; Berry and Hare 2004).

In the intervening years, following the discovery of superoxide radical and hydrogen peroxide production by mitochondria, the theoretical framework for the role of free radicals in Biology and Medicine was developed in seminal papers mainly by Briton Chance and associates (Chance et al. 1979; Naqui et al. 1986) and Trevor Slater (Slater 1984), and a book on the topic has, subsequently, been published (Halliwell and Gutteridge 1989). Even if nowadays causal relationships are difficult to establish between free radicals and disease it is clear that at least free radicals propagate the damage as part of a cycle leading to cell dysfunction and pathology. The cellular production of free radicals was then considered "the dark side of metabolism" (Valentine et al. 1998): if not neutralized, free radicals can oxidize double bonds in polyunsaturated membrane lipids, can react with nuclear and mitochondrial DNA, yielding strand breaks and also promote protein oxidation, resulting in increased degradation or loss of activity.

In view of the threats imposed by oxygen utilization, life in an aerobic environment, then, required antioxidant strategies to prevent and repair oxidative damage to vital cell components, as a consequence of an increase in the production of reactive oxygen species. In addition to SOD, other proteins have evolved to specifically and catalytically remove the reactive byproducts of oxidative metabolism, including, among several others, catalase and glutathione peroxidase. Of note, the cellular localization of these enzymatic activities are complementary to the major sources of superoxide radical and hydrogen peroxide, which points to regulated steady-state levels of these reactive species (Chance et al. 1979; Boveris 1998; Cadenas and Davies 2000). Although less specifically, evolutionary attempts to control free radicals and oxidants also include low molecular weight molecules, notably vitamin E, vitamin C, glutathione and uric acid. Conversely to these low molecular weight antioxidants, however, the free radical removal by enzymatic catalysis is not only more efficient, but occurs with formation of stable products without collateral production of reactive species and damage. That is, considering that radicals tend to be one-electron reactants, its chemical stabilization by low molecular weight antioxidants (e.g. vitamin E) via one-electron (or H atom) transfer often leads to a radical derived from antioxidant, which in turn, may induce deleterious oxidation (Stocker 1999). The overall understanding of antioxidant strategy is considered redundant and encompasses interconnected levels, from prevention of free radical formation, to interception/scavenging and repair and also to adaptation, as reviewed by Cadenas (Cadenas 1997).

As a corollary of these discoveries concerning the occurrence of radical/oxidants, on the one hand, and antioxidants, on the other, Helmut Sies formulated the concept of *oxidative stress*: an imbalance in the dynamic equilibrium between oxidants and antioxidants that favors the formers, potentially leading to damage (Sies 1985).

> A field of inquiry may be said to have come of age when conclusions initially viewed as remarkable or even unbelievable are accepted as commonplace. Study of the biology of free radicals, and the defenses thereto, has now reached this happy state of maturity (Fridovich 1997)

2.2 Free Radicals and Oxidants as Biological Messengers: From Oxidative Damage to Redox Signalling

Additionally to noxious oxidative challenges, free radical and oxidant-mediated processes were shown to modulate physiological pathways. During the 1980-90s, accumulating evidence has indicated that free radicals and oxidants play an essential role in cellular signalling and redox regulation of metabolic process. The post-translational modification of regulatory proteins that use redox chemistry is now accepted to transduce an oxidant signal into a biological response. The oxidative modification of critical groups in proteins constitutes a signal transduction mechanism, coupling the redox state of the protein to its function and activity. Overall, the cellular redox signalling is involved in cell growth and differentiation, adaptation and cell death with significance for physiological and pathological processes (Forman et al. 2002; Droge 2006; Janssen-Heininger et al. 2008).

Mechanistically, the thiol group of cysteine residues in proteins has been identified as one of the most sensitive sites for protein modification (Lipton et al. 1993), and is regarded as a major redox sensor that by interacting with a variety of oxidants form an array of potentially reversible modifications, including intra and inter disulfides, nitroso- and glutathionylated derivatives (Eaton 2006) (Klatt and Lamas 2000; Winterbourn and Hampton 2008).

Current evidences suggest that hydrogen peroxide and superoxide radical are components of signal transduction mechanisms for its intracellular production can be elicited by ligand-receptor, such as cytokine receptors, receptor tyrosine and serine/threonine kinases, G-protein-coupled receptors and ion channel-linked receptors, in response to several stimuli, including, among several others, angiotensin II, cytokines, glutamate, epidermal growth factor (EGF), vascular endothelial growth factor (VEGF), TNF-α, platelet-derived growth factor (PDGF) (Thannickal and Fanburg 2000).

Not only kinases (notably the MAPK family) and phosphatases can be modified by oxidants but also metabolic enzymes, cytoskeletal elements, cell-cycle and cell death control factors and a large collection of redox-sensitive targets have been identified, including universal transcription factors such as NF-*k*B, AP-1, CREB, Nrf2 and p53, thus implying free radicals and oxidants in the redox regulation of gene expression (Forman et al. 2002; Meng et al. 2002; Liu et al. 2005; Pantano et al. 2006; Humphries et al. 2007; Han et al. 2008; Liu et al. 2008; Monteiro et al. 2008; Trachootham et al. 2008).

At a subcellular level, mitochondria is not only a major source of free radicals but also a critical target implicated in redox signalling with other compartments, with impact in cell survival and death pathways (Cadenas 2004).

Redox signalling in angiogenesis is a relevant example because the process of new blood vessel growth is not only involved in physiological process, such as embryonic development and wound repair, but also contributes to several pathologies, including cancer, diabetic retinopathy, inflammation and atherosclerosis (Ushio-Fukai and Alexander 2004) (see Chapter 5. Considering that angiogenesis is predictably regulated by growth factors (e.g., EGF, VEGF) that elicit reactive species formation and that downstream targets (e.g., NF-*k*B) are targets (among others, such as matrix metalloproteinases) for redox modification by ROS, a role for these reactive species has been proposed in triggering angiogenic responses to ischemic-induced damage (Maulik and Das 2002).

An aspect of increasing relevance is the potential cross-talk between redox-dependent posttranslational modification and protein phosphorylation as coordinated mechanisms in signalling (Chiarugi and Buricchi 2007).

However, it should be clearly seen that the chemical reactivity of ROS is quite distinct and implications in signalling are expected. For instance, hydrogen peroxide exhibits longer life-times than superoxide radical and, conversely to the later, is not charged, diffusing through cell membranes (Boveris 1998). Regarding hydroxyl radical, its indiscriminate reactivity, as inferred from the extremely high reduction potential (strong oxidant), and short life-time ($t1/2 = 10^{-9}s$), excludes it from qualifying as a signalling molecule. Because of its stability and reactivity, hydrogen peroxide can have more subtle and specific actions on cells being, apparently, more suited to function as a cellular messenger. On the other hand, it has been suggested that a higher specificity of superoxide radical anion may be achieved providing that positively charged residues surround the site of action in the target molecule (Barrett et al. 1999). Similarly, due to high electrostatic attraction, it is well known that superoxide radical oxidizes iron-sulfur clusters in proteins at rates close to diffusion-controlled, releasing iron (McCord 2000). Conversely to superoxide radical, hydrogen peroxide (a non-radical molecule), by virtue of a two-electron oxidation, will oxidize thiols, for instance during hydroperoxide removal by thiol peroxidases, affecting the cellular redox environment and, thus, may set the conditions for redox-sensitive switches to act in one direction or the other (Saran et al. 1998).

The discovery of nitric oxide (•NO) in mammal cells added a new dimension to the notion of free radicals as biological messengers. Nitric oxide, synthesized by a family of nitric oxide synthases, is an ubiquitous gaseous radical and intercellular messenger involved in the regulation of major organs and systems, including immune, cardiovascular and nervous system ((Moncada et al. 1991; Bredt and Snyder 1994). Due to low molecular weight and hydrophobic properties nitric oxide easily permeates cell membranes and may diffuse a few cell diameters from its site of synthesis to neighbouring cells, a property that, in the brain, implicates the integration of the activity of neurons in a volume of tissue regardless of synaptic connections among them (Ledo et al. 2005). In signalling, the heme group in soluble guanylate ciclase (sGC), which is activated by nitric oxide, producing cGMP, has been identified as a major target (Ignarro 1991). However, recent findings point to the involvement of nitric oxide in signalling pathways independent of cGMP production (Boehning and Snyder 2003), in particular those mediated by its redox chemistry with thiol groups in proteins and glutathione, likely involving transition-metal catalysis and interconversion among different redox forms of nitric oxide (e.g. NO^+, the one-electron oxidation derivative). The formation of S-nitrosothiol proteins, occurring at a single critical cysteine residue has been described in many proteins and may regulate the function of transcription factors, receptors, ion channels, G-proteins and several enzymes (including protein kinases, phosphatases and caspases), structural proteins, etc. ((Stamler et al. 2001) Moreover, the competition between oxygen and nitric oxide for the binuclear center in cytochrome c oxidase points to the regulation of mitochondrial respiration by nitric oxide (reviewed in (Cooper et al. 2008).

The discovery of nitric oxide brought also a new collection of reactive species to be added to those derived from oxygen. In fact, nitric oxide is endowed with a low reactivity similar to that of oxygen but under relatively high oxygen tensions, via autoxidation, forms NO_2 radicals and N_2O_3, species with nitrating and nitrosating properties. Moreover, the diffusion-controlled reaction of nitric oxide with superoxide radical yields the potent oxidant and nitrating ion, peroxynitrite, ($ONOO^-$) (Radi 2004). When considering the biological activity of nitric oxide and superoxide radical, it is also interesting to note that, in view of the diffusion properties of the former, when both species are synthesized within a few cell diameters from each other they will combine at diffusion-limited rates to form peroxynitrite, outcompeting endogenous SOD for dismutation. That is, due to its low reactivity, direct actions of nitric oxide rely largely on binding to heme groups in proteins, but indirect reactions, via formation of nitric oxide-derivatives by autoxidation and reaction with superoxide radicals, the so-called *reactive nitrogen species* (RNS), give rise to a rich chemistry in the biological milieu, encompassing the oxidation, nitration and nitrosation of DNA bases, proteins and lipids with impact in pathophysiology. Thus, the term *nitrosative stress* and *nitroxidative stress* (Lancaster 2006) were introduced to account for the N-based radicals and oxidants.

As in the case of ROS, it is of note that RNS are a group of related molecules with individually distinct chemical and biological properties. Therefore the terms *oxidative and nitrosative stresses* implicate an array of distinct chemical reactions, including hydroxylations, peroxidations, oxidation of sulfhydryls to disulfides, sulfenic, sulfinic and sulfonic acids, carbonylations, nitrosations, nitrosylations, and nitrations.

In addition to a general involvement in nitrosative stress, the selective modification of cellular pathways by RNS, other than nitric oxide, has also been suggested. For instance, upon site-specific tyrosine nitration of cytochrome c (mediated either by peroxynitrite or by NO_2 radical) the protein gains a strong peroxidase function (Cassina et al. 2000). This modification may have critical biological impact because, nitrated citochrome c-dependent oxidation of cardiolipin in the inner mitochondrial membrane facilitates cytochrome c release to the cytosol with the well-known implications for activation of pro-caspase 9 and programmed cell death (Kagan et al. 2005; Orrenius and Zhivotovsky 2005). Likewise, nitrated fibrinogen accelerates clot formation ((Vadseth et al. 2004) and activation and translocation of protein kinase C*epsilon* occurs upon nitration (Balafanova et al. 2002). Additionally to the gain-of-function modification by nitration, from the viewpoint of loss-of-function upon protein nitration, an extremely relevant example is the mitochondrial matrix SOD isoform (MnSOD) that is inactivated via nitration by peroxynitrite (Radi 2004). The decreased removal of mitochondrial superoxide radical by nitrated MnSOD establishes a catalytic cycle leading to increased mitochondrial peroxynitrite which, in turn, may trigger cell death pathways via, for instance, oxidation-dependent opening of permeability transition pore (Radi et al. 2002).

Questions still persist as to how diffusible, chemically reactive species and short-lived ROS and RNS that kill cells meet the conditions for specific interactions in signalling pathways. Thus, current challenges regarding ROS and RNS as cell regulators are principally related with the specificity of their actions and, in a way reminiscent of the phosphorylation:dephosphorylation reactions, the reversibility of modifications they induce (Stamler et al. 1997; Stamler et al. 2001; Nathan 2003; D'Autreaux and Toledano 2007; Winterbourn 2008; Winterbourn 2008).

2.3 The Concept of Oxidative Stress Updated

The concept of oxidative stress has been seminal in providing a conceptual framework for planning experiments to study the significance of free radicals, oxidants and antioxidants in the context of human health. Notable progress has been achieved on the involvement of free radicals in disease and, significantly, epidemiological studies strongly suggest that an improved antioxidant status is associated with a reduced risk of diseases, such as atherosclerosis.

More recently, as discussed above, the precise regulation of free radicals and antioxidants production coupled to the identification of cellular pathways that use free radicals and oxidants as mediators implicated these species in the redox regulation of cell functions and gene expression. Individual ROS and RNS can play distinctive roles (Devadas et al. 2002) and in the case of hydrogen peroxide, increased steady-state levels have been involved in adaptation, proliferation, differentiation, apoptosis and necrosis (Antunes and Cadenas 2001; Desaint et al. 2004; Forman 2007). Interestingly, the Keap/Nrf2 pathway has been highlighted as a close fit to a receptor for radicals and oxidants receptor in mammals (D'Autreaux and Toledano 2007). This pathway is prominent in oxidative and environmental stress responses with target genes including phase II xenobiotic enzymes and antioxidants, being a target for hydrogen peroxide, lipid oxidation products, nitric oxide and several dietary polyphenols with strong antioxidant activity *in vitro* (Dinkova-Kostova et al. 2005). That is, the appreciation of the involvement of ROS and RNS in discreet redox pathways suggest that specific mechanisms have evolved for ROS/RNS signalling and, consequently, argues against the concept of "balance" in which distinct biological pathways respond equally to decreased pro-oxidants and increased antioxidants (D'Autreaux and Toledano 2007).

The very concept of "imbalance" between oxidants and antioxidants led to the search for biomarkers of oxidative damage inflicted to biomolecules and, in a complementary way, to the implementation of "antioxidant therapies" and interventions to "re-balance" the disrupted equilibrium. In both cases, however, the results are ambiguous and conflicting and, moreover, molecules with strong antioxidant activity *in vitro* (notably vitamin E) revealed effects *in vivo* not related with antioxidant activity (Azzi et al. 2004). The notion of a systemic "antioxidant" seems, therefore, barely supported in the light of "oxidative stress" as a "*global*" imbalance.

Attempts to quantify oxidative stress indicated that major thiol/disulfide couples, including GSH/GSSG, cysteine/cystine (Cys/CySS), thioredoxin1 reduced/ thioredoxin-1 oxidized are not in redox equilibrium and, moreover, respond differently to chemical stress, suggesting independent control of redox-sensitive pathways and that a global balance between pro-oxidant and antioxidant systems provides a limited view of oxidative stress (Jones 2002; Jones 2006).

The several lines of reasoning mentioned above led Jones (2006) to redefine oxidative stress as a *disruption of redox signalling and control* to emphasize discreet and compartimentalized cellular redox circuits.

This updated notion of oxidative stress offers the conceptual framework to study the molecular function of free radicals, oxidants and antioxidants in particular situations with impact in both health and disease. Of particular relevance can be envisaged the activity of the so-called dietary polyphenol antioxidants, such as flavonoids, which have been proposed to exert health benefits in a multitude of diseases states, such as cardiovascular disease, cancer, neurodegenerative diseases, as well as in the metabolic syndrome-associated disorders (see Chapter 8). It has become clear, however, that flavonoids are poorly bioavailable and extensively metabolized by gut microflora and by enzymatic conjugation in different tissues, including endothelial cells and liver. The metabolization occurs in a way that the chemical features responsible for H-donating and antioxidant activity are blocked by glucuronic acid, sulphate and methyl groups, suggesting that concentrations *in vivo* required for a meaningful antioxidant activity, outcompeting antioxidants such as ascorbate present at high micromolar concentration, can be hardly achieved (Walle 2004). On the other hand, several kinases (MAPK, protein kinase C, phosphoinositide 3-kinase, tyrosine kinases, Akt/PKB) and transcription factors (NF-*k*B, AP-1, Nrf2) have been shown to have their activity modified by flavonoids, which supports a role for flavonoids as modulators of cell signalling (Williams et al. 2004) (Frade et al. 2005).

The metabolic syndrome encompasses a wide collection of abnormalities associated with a high risk of developing type 2 diabetes and cardiovascular disease. Polyphenols, in particular the phytoalexin resveratrol has recently attracted much interest due to its capacity to, beyond direct antioxidant activities and at very low concentrations likely found *in vivo,* modulate several metabolic pathways in a way that may have a positive impact in human health (Das and Das 2007). One attractive idea is the ability of resveratrol to activate NAD+-dependent deacetylases sirtuins (e.g., SIRT1) and, thus, prevent metabolic dysfunctioning (Milne et al. 2007; Elliott and Jirousek 2008). SIRT1 is a key regulator of energy and metabolic homeostasis, and considering its beneficial effects on glucose homeostasis and insulin sensitivity, is emerging as a novel target for metabolic disease (Elliott and Jirousek 2008). The potential pharmacologic use of plant polyphenols, such as resveratrol, in the metabolic syndrome is a salient example that the new updated concept of oxidative stress may thus help project novel therapeutic approaches selectively directed to targets and disease conditions.

Acknowledgments Supported by FCT and FEDER, grant PTDCI/AGR-ALI/71262/2006.

References

Antunes F, Cadenas E. *Cellular titration of apoptosis with steady state concentrations of H2O2: submicromolar levels of H2O2 induce apoptosis through Fenton chemistry independent of the cellular thiol state.* Free Radic Biol Med. 2001; 30: 1008–18.

Azzi A, Davies KJ, et al. *Free radical biology. terminology and critical thinking*. FEBS Lett. 2004; 558: 3–6.

Babior BM, Kipnes RS, et al. *Biological defense mechanisms. The production by leukocytes of superoxide, a potential bactericidal agent.* J Clin Invest. 1973; 52: 741–744.

Balafanova Z, Bolli R, et al. *Nitric oxide (NO) induces nitration of protein kinase Cepsilon (PKCepsilon), facilitating PKCepsilon translocation via enhanced PKCepsilon -RACK2 interactions: a novel mechanism of no-triggered activation of PKCepsilon.* J Biol Chem. 2002; 277: 15021–7.

Barrett WC, DeGnore JP, et al. *Roles of superoxide radical anion in signal transduction mediated by reversible regulation of protein-tyrosine phosphatase 1B.* J Biol Chem. 1999; 274: 34543–6.

Berry CE, Hare JM. *Xanthine oxidoreductase and cardiovascular disease: molecular mechanisms and pathophysiological implications.* J Physiol. 2004; 555: 589–606.

Boehning D, Snyder SH. *Novel neural modulators.* Annu Rev Neurosci. 2003; 26: 105–31.

Boveris A. *Biochemistry of free radicals: from electrons to tissues.* Medicina (B Aires). 1998; 58: 350–6.

Boveris A, Cadenas E. *Mitochondrial production of superoxide anions and its relationship to the antimycin insensitive respiration.* FEBS Lett. 1975; 54: 311–4.

Boveris A, Chance B. *The mitochondrial generation of hydrogen peroxide. General properties and effect of hyperbaric oxygen.* Biochem J. 1973; 134: 707–16.

Boveris A, Oshino N, et al. *The cellular production of hydrogen peroxide.* Biochem J. 1972; 128: 617–30.

Bredt DS, Snyder SH. *Nitric oxide: a physiologic messenger molecule.* Annu Rev Biochem. 1994; 63: 175–95.

Cadenas E. *Basic mechanisms of antioxidant activity.* Biofactors. 1997; 6: 391–7.

Cadenas E. *Mitochondrial free radical production and cell signaling.* Mol Aspects Med. 2004; 25: 17–26.

Cadenas E, Davies KJ. *Mitochondrial free radical generation, oxidative stress, and aging.* Free Radic Biol Med. 2000; 29: 222–30.

Cassina AM, Hodara R, et al. *Cytochrome c nitration by peroxynitrite.* J Biol Chem. 2000; 275: 21409–15.

Chance B, Sies H, et al. *Hydroperoxide metabolism in mammalian organs.* Physiol Rev. 1979; 59: 527–605.

Chiarugi P, Buricchi F. *Protein tyrosine phosphorylation and reversible oxidation: two cross-talking posttranslation modifications.* Antioxid Redox Signal. 2007; 9: 1–24.

Cooper C, Mason M, et al. *A dynamic model of nitric oxide inhibition of mitochodnrial cytochrome c oxidase.* Biochim Biophys Acta. 2008; 1777: 867–876.

Cross CE, Halliwell B, et al. *Oxygen radicals and human disease.* Ann Intern Med. 1987; 107: 526–45.

D'Autreaux B, Toledano MB. *ROS as signalling molecules: mechanisms that generate specificity in ROS homeostasis.* Nat Rev Mol Cell Biol. 2007; 8: 813–24.

Das S, Das DK. *Resveratrol: a therapeutic promise for cardiovascular diseases.* Recent Patents Cardiovasc Drug Discov. 2007; 2: 133–8.

Desaint S, Luriau S, et al. *Mammalian antioxidant defenses are not inducible by H2O2.* J Biol Chem. 2004; 279: 31157–63.

Devadas S, Zaritskaya L, et al. *Discrete generation of superoxide and hydrogen peroxide by T cell receptor stimulation: selective regulation of mitogen-activated protein kinase activation and fas ligand expression.* J Exp Med. 2002; 195: 59–70.

Dinkova-Kostova AT, Holtzclaw WD, et al. *The role of Keap1 in cellular protective responses.* Chem Res Toxicol. 2005; 18: 1779–91.

Droge W. *Redox regulation in anabolic and catabolic processes.* Curr Opin Clin Nutr Metab Care. 2006; 9: 190–5.

Eaton P. *Protein thiol oxidation in health and disease: techniques for measuring disulfides and related modifications in complex protein mixtures.* Free Radic Biol Med. 2006; 40: 1889–99.

Elliott PJ, Jirousek M. *Sirtuins: novel targets for metabolic disease.* Curr Opin Investig Drugs. 2008; 9: 371–8.

Forman HJ. *Use and abuse of exogenous H2O2 in studies of signal transduction.* Free Radic Biol Med. 2007; 42: 926–32.

Forman HJ, Torres M, et al. *Redox signaling.* Mol Cell Biochem. 2002; 234–235: 49–62.

Frade JG, Ferreira RM, et al. *Mechanisms of neuroprotection by polyphenols.* Curr Med Chem – Central nervous system agents. 2005; 5: 307–318.

Fridovich I. *Superoxide anion radical (O2-.), superoxide dismutases, and related matters.* J Biol Chem. 1997; 272: 18515–7.

Fridovich I. *The biology of oxygen radicals.* Science. 1978; 201: 875–80.

Gerschman R, Gilbert DL, et al. *Oxygen poisoning and x-irradiation: a mechanism in common.* Science. 1954; 119: 623–6.

Halliwell B, Gutteridge JMC. *Free Radicals in Biology and Medicine*. Oxford, Clarendon Press. 1989.
Han ES, Muller FL, et al. *The in vivo gene expression signature of oxidative stress*. Physiol Genomics. 2008; 34: 112–26.
Harman D. *Aging: a theory based on free radical and radiation chemistry*. J Gerontol. 1956; 11: 298–300.
Humphries KM, Pennypacker JK, et al. *Redox regulation of cAMP-dependent protein kinase signaling: kinase versus phosphatase inactivation*. J Biol Chem. 2007; 282: 22072–9.
Ignarro LJ. *Signal transduction mechanisms involving nitric oxide*. Biochem Pharmacol. 1991; 41: 485–90.
Janssen-Heininger YM, Mossman BT, et al. *Redox-based regulation of signal transduction: principles, pitfalls, and promises*. Free Radic Biol Med. 2008; 45: 1–17.
Jones DP. *Redox potential of GSH/GSSG couple: assay and biological significance*. Methods Enzymol. 2002; 348: 93–112.
Jones DP. *Redefining oxidative stress*. Antioxid Redox Signal. 2006; 8: 1865–79.
Kagan VE, Tyurin VA, et al. *Cytochrome c acts as a cardiolipin oxygenase required for release of proapoptotic factors*. Nat Chem Biol. 2005; 1: 223–32.
Klatt P, Lamas S. *Regulation of protein function by S-glutathiolation in response to oxidative and nitrosative stress*. Eur J Biochem. 2000; 267: 4928–44.
Knowles PF, Gibson JF, et al. *Electron-spin-resonance evidence for enzymic reduction of oxygen to a free radical, the superoxide ion*. Biochem J. 1969; 111: 53–58.
Lancaster JR, Jr. *Nitroxidative, nitrosative, and nitrative stress: kinetic predictions of reactive nitrogen species chemistry under biological conditions*. Chem Res Toxicol. 2006; 19: 1160–74.
Ledo A, Barbosa RM, et al. *Concentration dynamics of nitric oxide in rat hippocampal subregions evoked by stimulation of the NMDA glutamate receptor*. Proc Natl Acad Sci U S A. 2005; 102: 17483–8.
Lipton SA, Choi YB, et al. *A redox-based mechanism for the neuroprotective and neurodestructive effects of nitric oxide and related nitroso-compounds*. Nature. 1993; 364: 626–32.
Liu B, Chen Y, et al. *ROS and p53: a versatile partnership*. Free Radic Biol Med. 2008; 44: 1529–35.
Liu H, Colavitti R, et al. *Redox-dependent transcriptional regulation*. Circ Res. 2005; 97: 967–74.
Loschen G, Azzi A, et al.*Mitochondrial H2O2 formation at site II*. Hoppe Seylers Z Physiol Chem. 1973; 354: 791–4.
Loschen G, Azzi A, et al. *Superoxide radicals as precursors of mitochondrial hydrogen peroxide*. FEBS Lett. 1974; 42: 68–72.
Loschen G, Flohe L, et al. *Respiratory chain linked H(2)O(2) production in pigeon heart mitochondria*. FEBS Lett. 1971; 18: 261–264.
Maulik N, Das DK. *Redox signaling in vascular angiogenesis*. Free Radic Biol Med. 2002; 33: 1047–60.
McCord JM. *The evolution of free radicals and oxidative stress*. Am J Med. 2000; 108: 652–9.
McCord JM, Fridovich I. *An enzymatic function for erythrocuproin*. J Biol Chem. 1969; 244: 6049–6055.
Meng TC, Fukada T, et al. *Reversible oxidation and inactivation of protein tyrosine phosphatases in vivo*. Mol Cell. 2002; 9: 387–99.
Milne JC, Lambert PD, et al. *Small molecule activators of SIRT1 as therapeutics for the treatment of type 2 diabetes*. Nature. 2007; 450: 712–6.
Moncada S, Palmer RM, et al. *Nitric oxide: physiology, pathophysiology, and pharmacology*. Pharmacol Rev. 1991; 43: 109–42.
Monteiro HP, Arai RJ, et al. *Protein tyrosine phosphorylation and protein tyrosine nitration in redox signaling*. Antioxid Redox Signal. 2008; 10: 843–89.
Naqui A, Chance B, et al. *Reactive oxygen intermediates in biochemistry*. Annu Rev Biochem. 1986; 55: 137–66.

Nathan C. *Specificity of a third kind: reactive oxygen and nitrogen intermediates in cell signaling.* J Clin Invest. 2003; 111: 769–78.
Orrenius S, Zhivotovsky B. *Cardiolipin oxidation sets cytochrome c free.* Nat Chem Biol. 2005; 1: 188–9.
Pantano C, Reynaert NL, et al. *Redox-sensitive kinases of the nuclear factor-kappaB signaling pathway.* Antioxid Redox Signal. 2006; 8: 1791–806.
Pauling L. *The discovery of the superoxide radical.* Trends in Biochemical Sciences. 1979; 4: N270–N271.
Pryor WA. *Oxy-radicals and related species: their formation, lifetimes, and reactions.* Annu Rev Physiol. 1986; 48: 657–67.
Radi R. *Nitric oxide, oxidants, and protein tyrosine nitration.* Proc Natl Acad Sci U S A. 2004; 101: 4003–8.
Radi R, Cassina A, et al. *Peroxynitrite reactions and formation in mitochondria.* Free Radic Biol Med. 2002; 33: 1451–64.
Saran M, Michel C, et al. *Radical functions in vivo: a critical review of current concepts and hypotheses.* Z Naturforsch [C]. 1998; 53: 210–27.
Sies H (1985) *Oxidative stress.* Academic Press, London.
Slater TF. *Free-radical mechanisms in tissue injury.* Biochem J. 1984; 222: 1–15.
Stamler JS, Lamas S, et al. *Nitrosylation. the prototypic redox-based signaling mechanism.* Cell. 2001; 106: 675–83.
Stamler JS, Toone EJ, et al. *(S)NO signals: translocation, regulation, and a consensus motif.* Neuron. 1997; 18: 691–6.
Stocker R. *The ambivalence of vitamin E in atherogenesis.* Trends Biochem Sci. 1999; 24: 219–23.
Thannickal VJ, Fanburg BL. *Reactive oxygen species in cell signaling.* Am J Physiol Lung Cell Mol Physiol. 2000; 279: L1005–28.
Trachootham D, Lu W, et al. *Redox regulation of cell survival.* Antioxid Redox Signal. 2008; 10: 1343–74.
Ushio-Fukai M, Alexander RW. *Reactive oxygen species as mediators of angiogenesis signaling: role of NAD(P)H oxidase.* Mol Cell Biochem. 2004; 264: 85–97.
Vadseth C, Souza JM, et al. *Pro-thrombotic state induced by post-translational modification of fibrinogen by reactive nitrogen species.* J Biol Chem. 2004; 279: 8820–6.
Valentine, J.S, Wertz DL, et al. *The dark side of dioxygen biochemistry.* Curr Opin Chem Biol. 1998; 2: 253–62.
Walle T. *Absorption and metabolism of flavonoids.* Free Radic Biol Med. 2004; 36: 829–37.
Weisiger RA, Fridovich I. *Mitochondrial superoxide simutase. Site of synthesis and intramitochondrial localization.* J Biol Chem. 1973; 248: 4793–6.
Williams RJ., Spencer JP, et al. *Flavonoids: antioxidants or signalling molecules?* Free Radic Biol Med. 2004; 36: 838–49.
Winterbourn CC. *Reconciling the chemistry and biology of reactive oxygen species.* Nat Chem Biol. 2008; 4: 278–86.
Winterbourn, CC, Hampton MB. *Thiol chemistry and specificity in redox signaling.* Free Radic Biol Med. 2008.
Zweier, JL, Broderick R, et al. *Determination of the mechanism of free radical generation in human aortic endothelial cells exposed to anoxia and reoxygenation.* J Biol Chem. 1994; 269: 24156–62.

Chapter 3
Oxidative Stress in the Metabolic Syndrome

Conceição Calhau and Alejandro Santos

Contents

C. Calhau (✉)
Department of Biochemistry (U38/ FCT), Faculty of Medicine, University of Porto,
Al. Prof. Hernâni Monteiro, 4200-319 Porto, Portugal
e-mail: ccalhau@med.up.pt

R. Soares, C. Costa (eds.), *Oxidative Stress, Inflammation and Angiogenesis in the Metabolic Syndrome,* DOI 10.1007/978-1-4020-9701-0_3,

Abstract This chapter describes the relationships between oxidative stress and the metabolic syndrome. Evidence suggests that oxidative stress may be involved in the aetiology, pathogenesis, and development of the metabolic syndrome. Additionally an important question remains to be asked: should we focus on antioxidant supplementation for managing the progression of complications, or on earlier steps before the development of complications? We think that it is crucial to interfere in both steps to prevent and treat several diseases, including the metabolic syndrome. Furthermore, antioxidant therapy may not only be too late, but it may also miss a large fraction of the target, as nonoxidative pathways do also contribute to cell damage.

Keywords Adipocyte · Dyslipidaemia · Hypertension · Leptin · Oxidative stress

3.1 Oxidative Stress and Obesity

Type 2 diabetes, hyperlipidaemia, hypertension, and atherosclerosis have recently been defined as typical lifestyle-related diseases. The common background of these diseases is **obesity**. Obesity, especially visceral obesity, with a rising prevalence in the last decade, is the main starting point to metabolic syndrome. Although obesity is common among patients with metabolic syndrome, not all obese people have it (Furukawa et al. 2004; Hansel et al. 2004; Ford 2006; Cardona et al. 2008a; Skalicky et al. 2008). On the other hand, lean people can develop this syndrome as well (Sjogren et al. 2005).

We will start by discussing how excess fat affects adipocyte metabolism, and how can oxidative stress be involved in adipocyte dysfunction, to consider then consequences of this on several other tissues.

Obesity can promote the atherosclerotic process through influence on endothelial function (see next section) as well as through mechanisms of oxidative stress (Furukawa et al. 2004; Skalicky et al. 2008). Adipocytes produce a variety of biologically active molecules, collectively known as adipokines, including TNF-α, resistin, leptin, and adiponectin (Bays et al. 2008). Adipocyte dysfunction provokes a deregulated production of these adipocytokines, which participate in the pathogenesis of obesity-associated metabolic syndrome. As an example, increased production of TNF-α from accumulated fat contribute to the development of insulin resistance in obesity.

It has been well known that oxidative stress occurs more frequently in people with metabolic syndrome phenotype (for definition, see chapter I) than among those without (Furukawa et al. 2004; Ford 2006; Cardona et al. 2008a; Skalicky et al. 2008), although not all studies have these conclusions (Sjogren et al. 2005). It is important to know if oxidative stress is a cause or a consequence in metabolic syndrome complications, i.e. whether oxidative stress occurs at an early stage, preceding the appearance of complications, or whether it is merely a common consequence of the cell damage, reflecting the presence of complications (probably, both). Part of the controversy has to do with what happens first (which remains unclear). What are the causes and what are the effects? As an example, some believe metabolic syndrome (insulin resistance, hypertension or obesity) causes oxidative stress (Furukawa et al. 2004;), whereas others believe oxidative stress causes insulin resistance, hypertension, atherosclerosis, obesity, and so on, i.e., metabolic syndrome (Baynes and Thorpe 1999; Evans et al. 2003; Urakawa et al. 2003; Evans et al. 2005; Skalicky et al. 2008).

Oxidative stress in adipocyte seems to be responsible for the sub-clinical pro-inflammatory state often observed in visceral obesity (Bastard et al. 2006; de Ferranti and Mozaffarian 2008). In fact, Furukawa et al. (2004) found a good correlation between obesity and systemic oxidative stress. Additionally, they observed a higher expression of NADPH oxidase that is accompanied by a decrease of antioxidant enzymes. The mechanisms involved in elevation of oxidative stress and inflammatory burden seem to include increased production of superoxide anion via the NAD(P)H oxidase pathway. NADPH oxidase activity causes deregulated production of adipocytokines, as plasminogen activator inhibitor-1, IL-6, and monocyte chemotactic protein-1 (Furukawa et al. 2004).

Leptin, a hormone produced by adipocytes, acts on hypothalamic centers to regulate food intake and energy expenditure. Plasma concentrations of this hormone are proportional to the amount of adipose tissue. Leptin has an important role in obesity-induced oxidative stress. This hormone stimulates directly ROS production such as H_2O_2 and hydroxyl radical. Furthermore, leptin is a proinflammatory factor that stimulates the proliferation of monocytes and macrophages and the production of inflammatory cytokines. Indirectly, leptin stimulates production of inflammatory cytokines such as IL-6 and TNF-α, which increase NADPH oxidase activity and superoxide anion production. Finally, leptin reduces the activity of paraoxonase-1 (PON-1), an enzyme that protects against LDL oxidation (Vincent and Taylor 2006).

It has been clearly demonstrated that chronic imbalance of consumed *vs* expended calories causes increased storage of the excess energy in the form of adipocyte intracellular triglyceride stores. Interestingly, within several cell types (adipocytes, myocytes, endothelial or β-cells), excess energy substrate in the form of glucose or FFAs enter the citric acid cycle, resulting in the generation of excess mitochondrial NADH, and consequently ROS (Vincent and Taylor 2006; de Ferranti and Mozaffarian 2008; Martyn et al. 2008). When excessive NADH cannot be dissipated by oxidative phosphorylation, the mitochondrial proton gradient increases, and single electrons are transferred to molecular oxygen, forming superoxide anion (Stephens et al. 2008; Cardona et al. 2008b). As further described in the next section, ROS production possibly links obesity with insulin resistance in adipocyte and skeletal muscle cells as well.

In adipocyte, oxidative stress, as a result of several mitochondrial (metabolic substrates: hyperglycaemia, elevated FFAs, uncoupling proteins dysfunction, among others) or extramitochondrial inducers (NADPH oxidase, cytochrome P450, iNOS, among other enzymatic activities), seems to lead to endoplasmic reticulum stress, with UPR (Unfolded Protein Response), which converge in a systemic proinflammatory state (de Ferranti and Mozaffarian 2008; Gregor and Hotamisligil 2007).

There are several possible contributors to oxidative stress in obesity, especially considering a common dietary pattern, with hyperglycaemia, elevated FFAs, and an inadequate antioxidant intake. As a matter of fact, there are studies showing that obese individuals have a lower intake of phytochemical-rich foods, compared with nonobese (Vincent and Taylor 2006). Additionally, it is known that the activities of the major antioxidant enzymes may also be lower in obese individuals (Bełtowski et al. 2000).

Bougoulia et al. (2006) investigated the relationships between cytokines, proinflammatory products and oxidative stress (IL-6, CRP, isoprostane and glutathione peroxidase), as well as their relation to cardiovascular disease risk factor in obese women (with central obesity) and their possible modification by **weight reduction**. In that work, it was demonstrated that: (1) low values of glutathione peroxidase along with higher levels of isoprostane in obese women indicate defective protection mechanisms against atherosclerosis and oxidative stress; (2) weight reduction ameliorated those parameters. It has been shown that individuals who lose 5-10% of body weight may lose 30% of their visceral fat, which is an important finding in this context.

3.2 Oxidative Stress and Diabetes

Elevated levels of metabolic substrates (glucose and/or fatty acids) contribute to the diabetic phenotype. Both insulin resistance and decreased insulin secretion are major features of the pathophysiology of type 2 diabetes. Insulin resistance is now clearly considered a major risk factor for developing type 2 diabetes, and most often precedes the onset of this pathology by many years. Initially, insulin resistance seems to be compensated through hyperinsulinemia, and normal glucose tolerance is preserved. When the pancreas has no longer capacity for this increased production of insulin, does hyperglycaemia occur.

In this section, we propose that ROS/RNS and oxidative stress induced by elevations in glucose and possibly FFA levels play a key role in causing insulin resistance and beta-cell dysfunction by their ability to activate stress-sensitive signalling pathways.

3.2.1 Insulin Resistance (Peripheral Tissues: Adipocyte, Myocyte)

3.2.1.1 Historical Perspective

Insulin resistance is defined as the condition whereby the body's cells require more and more insulin to get the same effect on glucose uptake. Liver, brain, and red blood cells do not require insulin action for uptake of plasma glucose. Although insulin is not required by liver cells for glucose uptake, it regulates important functions in liver cells, such as gluconeogenesis. Thus, when we refer to insulin resistance, below, the target insulin-sensitive tissues of the periphery that are being considered mainly encompass adipose tissue and skeletal muscle.

Hyperglycaemia has been demonstrated to constitute a risk factor for development of diabetic complications (Martyn et al. 2008). There being no consensus in relation to the molecular mechanisms between hyperglycaemia and disease, it has been clearly demonstrated that chronic exposure to elevated glucose concentrations can cause damage in different type of cells (specially, β-cell, adipocyte and myocyte) by different mechanisms involving oxidative stress.

There are several theories about the origin of those complications, including AGE (advanced glycation end products) hypothesis (Vlassara 1997), aldose reductase hypothesis (Hotta 1995), reductive stress (pseudohypoxia) (Ido et al. 1997), true hypoxia (Cameron and Cotter 1997), carbonyl stress (Lyons and Jenkins 1997), oxidative stress (Baynes and Thorpe 1999), altered lipoprotein metabolism (Lyons and Jenkins 1997), increased PKC activity (Ishii et al. 1998) and altered growth factor (Pfeiffer and Schatz 1995) or cytokine (Sharma and Ziyadeh 1997) activities. In this long list it may happen that each hypothesis corresponds to a different point of view of a common pathogenic mechanism. It may also happen that different cells are sensitive to different mechanisms. However, these different theories overlap and intersect with one another. All of those pathways converge in oxidative stress: AGE formation can induce oxidative stress that can accelerate AGE formation, and so on.

The *AGE* (Advanced Glycation End product) *hypothesis* proposes that chronic accelerated chemical modification of proteins by reducing sugars in diabetes alters the structure and function of tissue proteins, contributing to pathophysiology. Converging with this, the *carbonyl stress hypothesis* means a generalized increase of reactive carbonyl precursors of AGEs, glycoxidation and lipoxidation products. Mechanistically, carbonyl stress, with consequent increased carbonyls, result from an imbalance between production and detoxification of these reactive groups (Baynes and Thorpe 1999). It must be noticed that this term includes both carbonyls derived from oxidative and nonoxidative pathways. If the carbonyls are derived exclusively from oxidative reactions, then the condition would be described as ***oxidative stress***. The distinction may not be completely academic. The carbonyl stress hypothesis is in fact a mixture of metabolic and chemical hypothesis: altered metabolism and compromised detoxification lead to increase of carbonyl formation, increased chemical modification of proteins, and then to oxidative stress and tissue damage, which converge on the development of complications. According to different authors, oxidative stress is a secondary event in the pathogenic process (Baynes and Thorpe 1999). Interestingly enough is that AGEs products originated by nonoxidative pathways may induce oxidative stress and apoptosis in cells, illustrating a possible cause-effect relationship between carbonyl stress and oxidative stress.

Finally, other authors propose that metabolic imbalances in different tissues, resulting from excess of glucose metabolism, induce a state of pseudohypoxia or a *reductive stress*, rather than oxidative stress, in tissues. Pseudohypoxia is characterized by an increase in the cellular $NADH/NAD^+$ ratio. In diabetes the redox shift is attributed not to oxygen deprivation, but to excessive metabolism of glucose through glycolysis and the polyol pathway or of lipids by β-oxidation (Baynes and Thorpe 1999).

3.2.1.2 Oxidative Stress as a Link Between Hyperglycaemia and Insulin Resistance

How do elevated glucose and possibly free fatty acids contribute to the pathophysiology of diabetes via oxidative stress?

There is some evidence that oxidative stress caused by hyperglycaemia and/or FFA occurs before complications of diabetes become clinically evident. Hyperglycaemia leads to elevated formation of ROS and/or RNS as a consequence of non-enzymatic protein glycation and glucose autoxidation. Hyperglycaemia does also induce enzymatic production of superoxide anion through activation of NAD(P)H oxidase. For example, a non-enzymatic protein glycation may depend on ROS (superoxide and hydroxyl) formation through metal-catalyzed glucose autoxidation (Fig. 3.1).

Glucose and fatty acid independent sources of oxidative stress include enzymes such as NADPH oxidase and xanthine oxidase, both able to convert molecular oxygen into superoxide anion. Mitochondrial superoxide dismutase (SOD) seems to be responsible for oxygen conversion into hydrogen peroxide. In the cytoplasm, catalase or glutathione peroxidase then detoxify this further forming water. Therefore, if there is increased NADPH activity, or reduced SOD or glutathione activity, ROS production will be increased. Curiously, detoxifying enzymes seem to be decreased in obese patients (Furukawa et al. 2004).

Intracellular glucose elevations stimulate the polyol pathway in which aldose reductase mediates conversion of glucose to sorbitol. Excess sorbitol causes oxidative damage and activates stress genes, as has been demonstrated in several animal models (Evans et al. 2002). Hyperglycaemia does also increase NADPH oxidase activity, and NADPH produces the superoxide anion. When glucose itself autooxidizes, it produces oxidants with reactivity similar to that of hydroxyl radical and superoxide anion.

Oxidative stress may also result from the metabolic impact of intracellular triglycerides. For example, by suppressing the mitochondrial adenine nucleotide transporter, excessive triglycerides may increase superoxide anion production within the mitochondrial chain, and this decreases intramitochondrial ADP levels. Electrons then accumulate within electron transport chain and react with adjacent oxygen to form superoxide anion.

Visceral adiposity linked with elevated FFAs and hyperglycaemia does also produce nitroxide radicals via PKC pathway.

Free radicals in cells and ROS-derived lipid peroxides directly damage proteins, lipids and nucleic acids. Since ROS are produced mainly in mitochondria, they are a primary target for ROS-mediated reactions, resulting in mitochondrial damage. It has been assumed that in different pathologies, oxidative stress is a result of mitochondrial dysfunction, in consequence of several different toxic agents.

Mitochondrial dysfunction may contribute to insulin resistance in connection with the role of uncoupling proteins (UCP) as oxidative stress protecting agents.

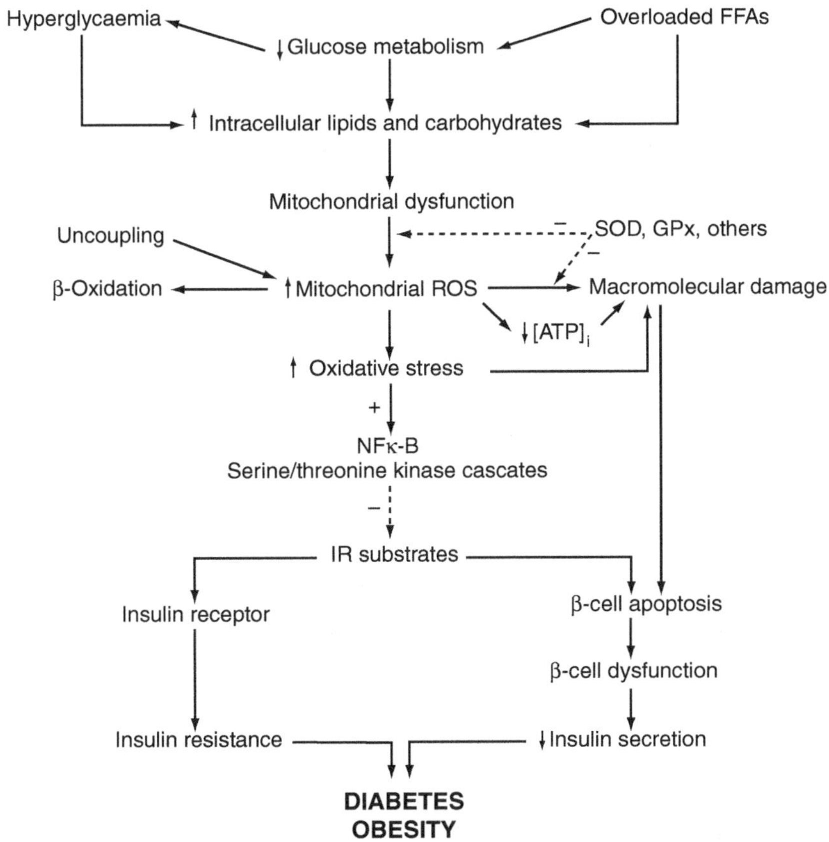

Fig. 3.1 Proposed link between oxidative stress, insulin resistance, beta-cell dysfunction and diabetes. Proposed general theory of how elevated glucose and possibly FFA levels contribute to the pathophysiology of diabetes via generation of oxidative stress, and the development of diabetic complications. The proposed sequence of events may also include other pathways not showed. FFAs, free fatty acids; GPx, glutathione peroxidase; ROS, reactive oxigen species; SOD, superoxide dismutase

Metabolic uncoupling refers to a state in which nutrient fuels are oxidized but the resultant energy is not linked to ATP synthesis but dissipated as heat. UCP1 is expressed in brown adipose tissue and shown to be an important thermogenic molecule. UCP2 and UCP3 can modulate cellular metabolism, although with a different tissue localization, suggestive of different physiological roles. Furthermore, according to Kim et al. (2008), overexpression of UCP2 or UCP3 lowers ROS production, stimulates the metabolic rate, and protects against weight gain and insulin resistance.

Molecular Mechanisms for Oxidative Stress-Induced Insulin Resistance

Given that both hyperglycaemia and increased FFA levels may result in ROS formation, it is important to understand how they converge in insulin resistance.

There is strong evidence to indicate that nuclear factor-κB (NF-κB), NH_2-terminal Jun kinases (JNK/SAPK), p38 mitogen-activated protein (MAP) kinase, and hexosamine pathways are stress-sensitive signalling systems activated by hyperglycaemia through oxidative stress. Thus, it has been accepted that activation of these pathways is linked not only to the development of the late complications of diabetes, but also to insulin resistance and β-cell dysfunction (Evans et al. 2003, 2005).

Additionally, strong evidence indicates that elevated FFA levels decrease insulin sensitivity and that pathway could be the link between obesity and diabetes (Randle et al. 1988). Elevated FFA levels have numerous adverse effects on mithocondrial function, as the uncoupling of oxidative phosphorylation, generating ROS. Furthermore, FFAs are not only able to induce oxidative stress but also to impair endogenous antioxidant defenses by reducing intracellular glutathione. It is probably through these pathways that FFAs could activate PKC-θ and consequently NK-κB (Slatter et al. 2000; Maassen et al. 2007). Additionally, it is known that fatty acids compete with glucose for metabolism, and fatty acid-derived metabolites, such as acetyl-CoA and citrate, inhibit insulin-stimulated glucose transport. Furthermore, impaired mitochondrial fatty acid oxidation, leading to the accumulation of fatty acid metabolites in muscle, is proposed to be a key factor in the development of insulin resistance in these cells. In good agreement with that, the results obtained by Cardona et al. (2008b) show an increase in oxidative stress after a fat overload, especially in patients with metabolic syndrome (Fig. 3.2).

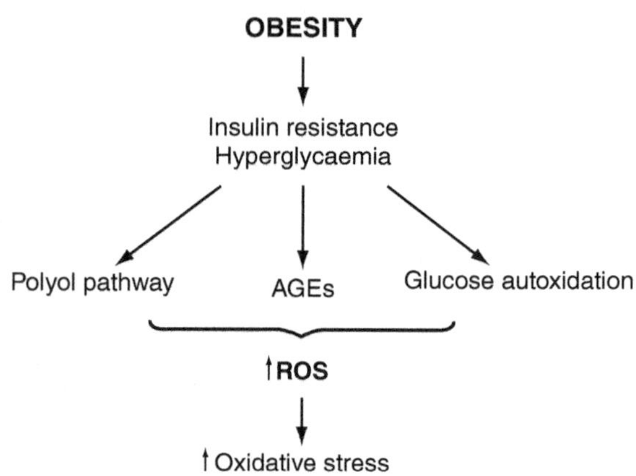

Fig. 3.2 Proposed link between obesity and oxidative stress. Proposed general theory of obesity as a cause of oxiadtive stress events. The proposed sequence of events may also include other pathways not showed. AGEs, advanced glycation end products; ROS, reactive oxigen species

3.2.2 β-Cell Dysfunction

Another target for oxidative stress damage is the β-cell. Recent data have implicated that β-cell dysfunction is the result of prolonged exposure to high glucose, elevated FFA levels, or a combination of both.

According with Evans hypothesis (Evans et al. 2003), hyperglycaemia induces deterioration of β-cell function probably through oxidative stress. In fact, there is evidence that oxidative stress leads to tissue damage. It has been described that ROS formation is a direct consequence of hyperglycaemia and of increased FFA levels. In addition to their ability to directly damage macromolecules (DNA, lipids and proteins), ROS can function as signalling molecules to activate a number of cellular stress-sensitive pathways that cause cellular damage and, thus, probably play a key direct role in the pathogenesis of late diabetic complications. These same pathways are linked to insulin resistance and decreased insulin secretion. β-Cells at high risk for oxidative damage have an increased sensitivity for apoptosis. These cells are particularly sensitive to ROS and RNS because they are poor in antioxidant enzymes such as SOD, glutathione peroxidase and catalase (Tiedge et al. 1997; Robertson et al. 2007).

3.2.2.1 Mitochondrial Dysfunction and Diabetes

Mitochondrial damage, as consequence of oxidative stress, includes decreased mitochondrial ATP synthesis and deregulation of intracellular lipid and calcium homeostasis, with important consequences on cell viability. In the β-cell, the main consequence for UCP2 (higher) activity, and thus for the lower cellular ATP, appears to be impairment of closure of plasma membrane ATP-dependent K^+ channels (Chan and Harper 2006; Fridlyand and Philipson 2006; Kim et al. 2008). Such modulation results in reduced glucose-stimulated insulin secretion.

3.2.2.2 Antioxidant Administration

Although oxidative stress is widely invoked as a pathogenic mechanism for diabetes, there is limited evidence yet that antioxidant vitamin and drug supplements provide protection against progression of this disease, either in human or in animal models (Lee et al. 2004; Steinhubl 2008). In contrast, interventions to decrease substrate (glucose and/or FFAs) concentrations have demonstrated protecting effects on the risk for, and progression of, diabetes. In spite of this, there are studies showing that plasma from diabetic patients present increased levels of end-products of oxidative damage (Nourooz-Zadeh et al. 1995; Borcea et al. 1999; Davì et al. 1999). Furthermore, although of short duration, several clinical trials show that vitamin E, vitamin C, or glutathione supplementation improves insulin sensitivity in insulin-resistant patients and/or patients with type 2 diabetes (Paolisso and Giugliano 1996; Evans and Goldfine 2000; Steinhubl 2008).

Altogether, we conclude from results obtained in different epidemiological studies that there seems to be no advantage for antioxidant supplementation.

3.3 Hypertension, Metabolic Syndrome and Oxidative Stress

Cardiovascular disease is now endemic worldwide and no longer limited to economically developed countries (Ezzati et al. 2002). About 7.6 million deaths (about 13,5% of the total) and 92 million Disability Adjusted Life Years (DALYs) (6,0%

of the total) worldwide were attributed to high blood pressure in 2001. High blood pressure was a major health issue in all world regions, and it accounted for more than a third of deaths and almost a fifth of DALYs in Europe and central Asia (Lawes et al. 2008).

High blood pressure (BP) values are one of the main metabolic syndrome components, and metabolic syndrome has been found in about 30–40% of hypertensives. Whether or not the presence of metabolic syndrome increases the hypertension-induced cardiovascular risk is a matter of debate. Moreover, criticisms about the existence of metabolic syndrome have been raised, and there are doubts concerning whether metabolic syndrome itself results in a higher risk than the sum effect of each of the components (Kahn et al. 2005). However, increasing evidence indicates that the clustering of metabolic and hemodynamic abnormalities characterizing the metabolic syndrome is associated with a prevalence of subclinical damage in a variety of organs, such as left ventricular hypertrophy, thickening or atherosclerotic plaques of carotid arteries, microalbuminuria and deranged renal function. This is clinically relevant since these markers of target organ damage are associated with an increased risk of cardiovascular fatal and nonfatal events. The contribution of the metabolic syndrome to target organ damage in hypertensives is presumably responsible for a substantial increase in cardiovascular fatal and nonfatal events (Cuspidi et al. 2008). Excessive production of reactive oxygen species (ROS), which exceeds antioxidant defence mechanisms, has been implicated in pathophysiological conditions that impact on the cardiovascular system. Hypertension is considered a state of oxidative stress that can contribute to the development of atherosclerosis (Romero 1999). Assessment of antioxidant activities and lipid peroxidation by-products in hypertensives indicates an excessive amount of ROS and a reduction of antioxidant defence activities in blood, as well as in several other cellular systems, including not only vascular wall cells (Orie et al. 1999), but also circulating cells (Yasunari et al. 2002). Antihypertensive treatment attenuated the increase in oxidative stress observed in hypertensive subjects, and extending treatment over time increased the beneficial impact of treatment (Saez et al. 2004).

3.4 Mechanisms Linking Hypertension and Oxidative Stress

It remains unclear whether elevated levels of free radicals initiate the development of hypertension, are a consequence of the disease process itself or both (Grossman 2008). Oxidative stress may add to the generation and/or maintenance of hypertension via several possible mechanisms. These include quenching of the vasodilator nitric oxide (NO) by ROS such as superoxide (McIntyre et al. 1999), depletion of tetrahydrobiopterin, an important NO synthase (NOS) cofactor (Vaziri et al. 2000), formation of vasoconstrictor lipid peroxidation products, such as F2-isoprostanes (Cracowski et al. 2002), as well as structural and functional changes within the vasculature (Zalba et al. 2001). These vascular alterations may be mediated in several ways, including direct damage to endothelial and vascular smooth muscle cells, effects on endothelial cell eicosanoid metabolism, altered redox state, increases in intracellular free calcium concentrations and stimulation of inflammatory and growth-signalling events (Chen et al. 2001; Ortiz et al. 2001; Zalba et al. 2001).

There are several possible sources of free radical production within the vasculature. These include NADPH oxidase, NOS, cyclo-oxygenases, lipoxygenases and xanthine oxidase, all of which are functional in endothelial cells (Fleming et al. 2001).

3.4.1 NAD(P)H Oxidase

Numerous studies have shown that the main source of free radicals in the vascular wall is nonphagocytic NAD(P)H oxidase, which uses NADH/NADPH as the electron donor to reduce molecular oxygen and produce $O_2^{\bullet -}$. Activation of this enzyme requires the assembly of both cytosolic (p47phox, p67phox or homologues) and membrane bound (gp91phox/Nox1/Nox4 and p22phox) subunits to form a functional enzyme complex. In the vasculature the NAD(P)H oxidase complex is at least partially pre-assembled, as a significant proportion of NAD(P)H oxidase subunits are colocalized within endothelial cells (Bayraktutan et al. 2000; Li et al. 2002). Activation of NAD(P)H oxidase is regulated by several vasoactive hormones, mechanical stimuli (shear stress and stretch) and growth factors (platelet-derived growth factor, transforming growth factor-β) (Lassegue et al. 2003). The NAD(P)H oxidase activation by angiotensin II pathway in vascular cells has been shown to involve protein kinase C, phospholipase D, c-Src and receptor tyrosine kinases (Seshiah et al. 2002). This is particularly relevant since it is known that: adipose tissue has the full molecular machinery required for local angiotensin II synthesis and angiotensin II stimulated signal transduction (Engeli et al. 2000); both levels of angiotensinogen and angiotensin II are increased in visceral obesity, a feature of metabolic syndrome (Harte et al. 2005). NAD(P)H oxidase is also found in the kidney (Touyz 2004) and plays a significant role in the renin–angiotensin system and establishing levels of angiotensin II. Angiotensin II is not only a potent vasoconstrictor, but is also a source of ROS and can stimulate production of endothelin-1. Slow intravenous infusion of angiotensin II in rats, which does not trigger an immediate pressor response, can increase F2-isoprostane levels and mean arterial pressure (Rodrigo et al. 2003). A recent study showed that urinary isoprostanes were no different in hypertensives *versus* controls, although hypertensives with renovascular disease had higher levels compared with both other groups (Minuz et al. 2002). Additionally, angiotensin II infusion has been shown to increase 20-hydroxyeicosatetraenoic acid (20-HETE) production in rat renal microvessels (Croft et al. 2000), that may contribute to vasoconstriction and elevation of BP.

In rats with hypertension caused by angiotensin II infusion, both NAD(P)H oxidase subunit expression and activity are increased (Fukui et al. 1997; Rajagopalan et al. 1996), whereas administration of a NAD(P)H oxidase inhibitor reduces vascular $O_2^{\bullet-}$ production and attenuates angiotensin II-induced increases in blood pressure (Rey et al. 2001). In mice lacking the NAD(P)H oxidase cytosolic subunit p47phox, the hypertensive response to angiotensin II is markedly attenuated, and these animals do not show the same increases in $O_2^{\bullet-}$ production and endothelial dysfunction observed in angiotensin II infused wild-type mice (Landmesser et al. 2002).

Most studies use exogenous angiotensin II to study its role in free radical generation in hypertension. Few studies show the role of the endogenous renin-angiotensin system in free radical production during hypertension. An increase in $O_2^{\bullet-}$ production from a gp91phox containing NAD(P)H oxidase has been linked to endothelial dysfunction in a model of renovascular hypertension (2-kidney 1-clip) that activates the renin-angiotensin system (Jung et al. 2004). It is noteworthy that atorvastatin decreases the expression of NAD(P)H oxidase subunits and upregulates catalase *in vivo* which could help explain the vasoprotective effects observed with the use of these drugs (Wassmann et al. 2002).

3.4.2 Nitric Oxide Synthase

Nitric oxide synthase (NOS) can also contribute to free radical production, since the three NOS isoforms have been shown to be susceptible to the uncoupling that leads to the formation of $O_2^{\bullet-}$ (instead of NO) under certain conditions (Andrew et al. 1999). For endothelial NOS, this process can be initiated *in vitro* through the absence of the co-factors L-arginine and tetrahydrobiopterin (Vasquez-Vivar et al. 1998). Importantly, uncoupling of endothelial NOS has been demonstrated in mice with DOCA-salt-induced hypertension (Landmesser et al. 2003). The critical step in this uncoupling seems to be oxidation of tetrahydrobiopterin by $ONOO^-$, reducing the bioavailability of this critical cofactor (Landmesser et al. 2003; Laursen et al. 2001). Treatment with tetrahydrobiopterin improves blood pressure in both DOCA-salt hypertension and spontaneously hypertensive rats (SHR) (Hong et al. 2001; Landmesser et al. 2003). Hemodynamic forces such as laminar and oscillatory shear stimulate an acute increase in NO production and upregulation of eNOS (Boo et al. 2003; Cai et al. 2004). Physical exercise increases vascular shear stress and is a key physiological mechanical activator of endothelial NO production and inducer of eNOS expression. These processes add to vascular changes linked with exercise training, including physiological remodelling, vasodilation, increased organ blood flow, angiogenesis and vascular protection (Kojda et al. 2005). Laminar shear increases the expression of the cytosolic copper/zinc-containing SOD and extracellular SOD, major scavengers of cytoplasmic and extracellular $O_2^{\bullet-}$, respectively (Inoue et al. 1996). Laminar flow also stimulates the expression and intracellular levels of glutathione (GSH) peroxidase, responsible for H_2O_2 scavenging (Harrison et al. 2006). In contrast, hypertension and atherosclerosis cause disturbed blood flow profiles (oscillatory shear) that promote oxidative stress and oxidative vascular damage. Oscillatory shear stress is associated with sustained $O_2^{\bullet-}$ production, which in the presence of NO, enhances $ONOO^-$ formation and protein nitration (Beckman et al. 1996).

3.4.3 Xanthine Oxidase

Xanthine oxidase is a metalloenzyme that catalyses the oxidation of hypoxanthine and xanthine to form $O_2^{\bullet-}$ and is present in the vascular endothelium. There is some evidence to suggest that xanthine oxidase-derived $O_2^{\bullet-}$ is involved in the endothelial dysfunction observed in hypertension. Spontaneously hypertensive rats (SHR) demonstrate high xanthine oxidase activity levels in the mesenteric microcirculation, and this is related with increased arteriolar tone (Suzuki et al. 1998). Transgenic rats overexpressing renin and angiotensinogen develop endothelial dysfunction; this has been linked with increased xanthine oxidase activity (Mervaala et al. 2001). Xanthine oxidase is involved in end-organ damage in hypertension. Both SHR and Dahl salt-sensitive rats exhibit increased xanthine oxidase activity in the kidney. In the SHR, long-term inhibition of xanthine oxidase with allopurinol reduced renal xanthine oxidase activity with no reduction in blood pressure, indicating that the increased renal ROS production was an effect of hypertension instead of a contributing factor. The knowledge that allopurinol can improve cardiac and renal hypertrophy in SHR while having a reduced effect on blood pressure supports the idea that xanthine oxidase has a role in hypertensive end-organ damage and not in the genesis of hypertension (Laakso et al. 1998).

3.4.4 Cytochrome P450 Enzymes

Also present within the vasculature are the cytochrome P450 enzymes (CYP450), which may also be significant sources of oxidative stress. During the metabolism of arachidonic acid by CYP450, an important cofactor is the NADPH oxidase system, which results in the production of superoxide ions (Sarkis et al. 2004). In addition to this, the CYP450 reaction cycle requires transfer of electrons from the central heme iron to the activated bound oxygen molecule, which can generate superoxide, hydrogen peroxide and hydroxyl radicals. *In vitro* studies have shown that the CYP450 2C9 isoform is a functionally significant source of ROS in coronary arteries (Fleming et al. 2001). The metabolism of arachidonic acid by the CYP450 4A isoform in smooth muscle cells leads to the formation of 20-hydroxyeicosatetraenoic acid (20-HETE), a potent vasoconstrictor. There is evidence to suggest that NO is capable of inhibiting the formation of 20-HETE by binding to the catalytic heme site in the CYP450 enzyme (Sun et al. 1998). There is a positive association between 20-HETE excretion and BP in women and a significant inverse association between 20-HETE excretion and endothelium-dependent vasodilation in both men and women (Ward et al. 2004). Additionally, 20-HETE excretion and markers of oxidative damage have been found to be associated (Ward et al. 2005). All this data sustain the idea that CYP450 is not only an important source of oxidative stress within the vessel wall, but also plays a role in endothelial dysfunction and hypertension.

3.4.5 Macronutrients, Obesity, Blood Pressure and Oxidative Stress

Chronic overnutrition has been linked with a significant increase in oxidative stress as reflected in indexes of lipid peroxidation, oxidative damage of amino acids and protein carbonylation. Interestingly, dietary restriction in these obese subjects reverted the oxidative damage. Different macronutrients provoke different patterns of ROS generation. The greater peak increase in ROS formation is caused by carbohydrates (glucose) and the least by protein. The ROS generation by mononuclear cells and polymorphonuclear leukocytes peaks 2 h after the ingestion of glucose and 1 h after the ingestion of lipids. Lipid ingestion causes a prolonged increase in lipid peroxidation (Dandona et al. 2001). High ingestion of carbohydrates, lipid or protein increases the formation of ROS by NAD(P)H oxidase (Mohanty et al. 2002). As previously referred obese individuals may have elevated levels of angiotensin II (Harte et al. 2005) which can increase NAD(P)H oxidase activity (Landmesser et al. 2002). This link between angiotensin II and oxidative stress could explain some of the beneficial effects of angiotensin receptor blockers such as valsartan or irbesartan, since the use of these drugs results in rapid and profound decreases in ROS formation (Dandona et al. 2003; Khan et al. 2001).

3.4.6 Antioxidant Enzymes

The biological effects of highly reactive ROS are controlled *in vivo* by a variety of non-enzymatic and enzymatic antioxidant mechanisms. Among the latter, SOD catalyses dismutation of the superoxide anion into H_2O_2, catalase detoxifies H_2O_2 and glutathione peroxidase (GPX) detoxifies H_2O_2 and converts lipid hydroperoxides into nontoxic alcohols. Hypertensive patients have significantly lower SOD and GPX activities when compared with controls. This could reflect inactivation of these free radical scavengers in a situation of increased oxidative stress or reduced production; in that way these patients may be more susceptible to damage. In hypertensive patients there is an imbalance between the damaging effects and the bioscavenging capability of superoxide and other reactive oxygen species. Such is the case of SOD activity, which is inversely correlated with blood pressure in these patients (Pedro-Botet et al. 2000). Additionally, endothelium-dependent relaxation is impaired in hypertensive patients owing to abnormal endothelial nitric oxide availability. Nitric oxide is rapidly inactivated by the superoxide anion (Beckman and Koppenol 1996). Since the protective scavenging function against superoxide is provided mainly by SOD, the low SOD levels seen in hypertensive patients could in turn be related to impaired endothelial function as a consequence of reduced superoxide anion removal (Lacy et al. 1998). The negative association found between blood pressure and SOD activity in hypertensive patients strengthens this hypothesis. Interestingly, intravenous SOD injection reduces arterial pressure in SHR, but not in normal rats (Nakazono et al. 1991). Patients with mild-to-moderate hypertension show low endogenous antioxidant enzyme activities. This may contribute to rendering the arterial wall more susceptible to oxidative injury in essential hypertension (Pedro-Botet et al. 2000). However, metabolic syndrome does not appear to increase the abnormalities in antioxidant mechanisms observed in hypertensives (Abdilla et al. 2007).

Several studies support that there is an increased level of oxidative stress in essential hypertensive patients (Kedziora-Kornatowska et al. 2004; Rodrigo et al. 2007; Simic et al. 2006). Even white-coat hypertensives have increased oxidative stress, namely through a reduction in paraoxonase 1 activity (Uzun et al. 2004). A recent study found that the increased GSSG/GSH ratio and high ROS-derived by-products in both blood and peripheral mononuclear cells observed in hypertensives were not enhanced by the presence of additional features of metabolic syndrome. Similarly, the reduction in the activity of antioxidant enzymes, both extracellular and cytoplasmic, observed in hypertensive patients was not affected by the presence of additional features of the metabolic syndrome. Neither the number of features nor the individual addition of each of them, abdominal obesity, low HDL, hypertriglyceridemia or fasting hyperglycaemia, contributed to the oxidative stress abnormalities observed in patients whose only symptom was arterial hypertension (Abdilla et al. 2007).

3.4.7 Antioxidants and Blood Pressure

The evidence from many clinical studies supporting antioxidant supplementation to lower blood pressure is limited. A study performed in patients with type 2 diabetes, supplemented with 500 mg/day RRR-α-tocopherol, 500 mg/day mixed tocopherols (60% γ-tocopherol) or placebo for 6 weeks, showed that treatment with either α- or mixed tocopherols significantly increased blood pressure, pulse pressure and heart rate (Ward et al. 2007). Epidemiological data support a protective role of vitamin E against cardiovascular disease (Meydani 2004). However, a recent meta-analysis has highlighted an increase in all-cause mortality following high-dose vitamin E supplementation (Miller et al. 2005). In the SU.VI.MAX study supplementing with 120 mg vitamin C, 30 mg vitamin E, 6 mg β-carotene, 100 mg selenium (in the form of selenium-enriched yeast) and 20 mg zinc (as gluconate) for 6.5 years did not demonstrate any beneficial effect upon the 6.5-year risk of hypertension (Czernichow et al. 2005). A similar result was obtained by the Heart Protection Collaborative Group; they did not observe any improvement in blood pressure after 5 years of treatment with a combination of ascorbic acid, synthetic vitamin E and β-carotene in high risk cardiovascular patients (MRC/BHF Heart Protection Study of antioxidant vitamin supplementation in 20,536 high-risk individuals: a randomised placebo-controlled trial 2002).

Smaller studies showed that supplementing with vitamins C and E reduces blood pressure (Rodrigo et al. 2008), improves arterial stiffness and endothelial function in essential hypertensive patients (Plantinga et al. 2007).

A recent study showed that ascorbic acid decreases the binding affinity of the AT_1 receptor to angiotensin II, this could be a mechanistic explanation to the blood pressure lowering effects reported in studies with vitamin C supplementation (Leclerc et al. 2008).

Many blood pressure lowering agents reduce oxidative stress, in spite of the fact that some of those agents have no known antioxidant activity per se; this situation favours the idea that lowering blood pressure rather than the agents used reduces oxidative stress. Therefore, it seems that oxidative stress is not the cause, but, on the contrary, a consequence of hypertension (Grossman 2008).

3.5 Dyslipidaemia

The metabolic syndrome and type 2 diabetes are characterised by a constellation of risk factors which act multiplicatively to promote cardiovascular disease. In particular, the atherogenic dyslipidaemic profile associated with these conditions, specifically mild to marked elevation of triglyceride-rich lipoprotein (very low-density lipoproteins [VLDL] and VLDL remnants) concentrations, an increase in small, dense LDL [low-density lipoproteins] and apolipoprotein B (apoB), and low levels of HDL-C [high-density lipoproteins - cholesterol] appears to play a prominent role (Adiels et al. 2006; Ginsberg et al. 2006). This dyslipidaemia is also characterised by a spectrum of qualitative lipid abnormalities reflecting perturbations in the structure, metabolism and biological activities of both atherogenic lipoproteins containing apoB (VLDL, intermediate-density lipoprotein [IDL] and LDL), and anti-atherogenic HDL containing apoA-I and/or apoA-II. These anomalies, which include functionally defective HDL, glycation of apoB and apoA 1 and a dense LDL phenotype, are closely associated with increased oxidative stress and endothelial dysfunction, thereby reinforcing the pro-inflammatory nature of macrovascular atherosclerotic disease. The excess of atherogenic apoB-containing lipoproteins (VLDL, VLDL remnants, IDL and LDL) promotes enhanced arterial cholesterol deposition and accelerated progression of atherosclerotic disease (Adiels et al. 2008; Parthasarathy et al. 2008). Accumulation of adipose tissue, particularly in the visceral compartment, drives the dyslipidaemia, hyperglycaemia and hypertension typically associated with the metabolic syndrome and type 2 diabetes, and promotes the development of micro- and macrovascular atherosclerotic disease (Bays et al. 2008). However, dyslipidaemia is probably the major mediator of atherogenicity in the metabolic syndrome (Alexander et al. 2003) and is characteristic of insulin resistance (Grundy 1998). The "lipid triad" (high levels of plasma triglycerides, low levels of HDL-C, and the appearance of small, dense LDL (sdLDL) (Grundy 1998) frequently precedes type 2 diabetes by several years, indicating that the disturbance of lipid metabolism is an early event in the progress of cardiovascular complications of type 2 diabetes (Ginsberg et al. 2006). In fact, patients with insulin resistance both with and without type 2 diabetes display qualitatively similar lipid abnormalities (Ginsberg et al. 2006). It is now accepted that the different components of diabetic dyslipidaemia are not isolated abnormalities but are closely linked to each other metabolically and are mainly initiated by the hepatic overproduction of large triglyceride rich very low–density lipoproteins (VLDL1) (Adiels et al. 2008; Ginsberg et al. 2006).

3.5.1 Atherosclerosis – The Oxidative Hypothesis

Atherosclerosis is the most important manifestation of cardiovascular diseases. The part of lipoproteins in the development of atherosclerosis has been studied for more than 50 years. It is now uncontroversial that high levels of low density lipoprotein (LDL) and low levels of high density lipoprotein (HDL) are major contributors to the development and progression of cardiovascular diseases (Rosin 2007). Even if the biochemical processes that contributed to the development of early atherosclerotic lesions, the fatty streak lesions, are still under study. The LDL oxidation hypothesis was formulated in the 1980's to explain the formation of fatty streak lesions, thousands of publications have appeared to date on this topic providing evidence for the presence of oxidative processes in human disease (Parthasarathy et al. 2008; Rosin 2007). The possible steps involved in the development of early atherosclerotic lesion outlined by the hypothesis are: increased plasma and intimal LDL; oxidation of lipids associated with LDL; chemotactic recruitment of monocytes and their differentiation into macrophages; alteration of apolipoprotein B100 to a negatively charged form (oxidatively modified LDL) by lipid peroxidation by-products; uptake of oxidatively modified LDL by macrophages; retention of lipid loaded macrophages in the intima (Parthasarathy et al. 2008).

3.5.2 Oxidatively Modified LDL

The observation that *in vitro* incubation of macrophages with oxidized LDL (oxLDL) and not with native LDL led to intracellular cholesterol ester accumulation (Henriksen et al. 1983) was the basis to the formulation of the oxidation hypothesis of atherosclerosis. The oxidation of LDL is a complex process involving both protein and lipid oxidative changes and the formation of complex products. The decomposition of peroxidized lipids generates both free and core ketones and aldehydes that covalently change ε-amino groups of lysine residues of the protein moiety (Haberland et al. 1982; Haberland et al. 1984). The oxidative modification is not exclusive to LDL as other lipoproteins such as very low density lipoprotein (VLDL), beta very low density lipoprotein (β-VLDL), and even high density lipoprotein (HDL) can undergo similar oxidative changes (Shao et al. 2006; Young et al. 2003) changing their pro- or anti-atherosclerotic behaviour. The main message from these studies was that the oxidative modification of LDL might be the key determinant of lipid uptake by macrophages.

Most authors believe that "fully oxidized LDL" does not exist in the circulation, mainly because of the high antioxidant levels observed in blood. Furthermore, such highly oxidized particles would be rapidly removed by the liver via scavenger receptors (Van Berkel et al. 1991). In contrast, the presence of circulating minimally oxidized LDL has already been demonstrated. The oxidative modification suffered by these particles is not sufficient to cause changes recognized by scavenger receptors, therefore they remain longer in circulation (Avogaro et al. 1988). This oxidized LDL is only a small fraction of LDL ranging from 0.001% in healthy controls (Shoji et al. 2000) to approximately 5% in patients with acute coronary events (Holvoet et al. 1998). Since LDL is a substrate for oxidation, concentrations of oxidized LDL correlate with LDL concentrations, and therefore with the cholesterol within LDL. Also, concentrations of oxidized LDL depend on the sensitivity of LDL particles to oxidation; sdLDL contains smaller amounts of antioxidants and therefore they are more susceptible to oxidation. A higher prevalence of sdLDL has been associated with metabolic syndrome (Lamarche 1998).

3.5.3 VLDL Overproduction as a Key Feature of Metabolic Syndrome Dyslipidaemia

The formation of sdLDL is closely associated with insulin resistance and hypertriglyceridemia (Verges 2005) and the VLDL1-triglyceride level is the major predictor of LDL size in individuals with or without type 2 diabetes. The dyslipidaemia seen in insulin resistance and type 2 diabetes is related with excessive hepatic production of VLDL (Adiels et al. 2006) mainly the VLDL1 fraction (Adiels et al. 2005). On the other hand, hepatic secretion of VLDL2 is similar in insulin-resistant and insulin-sensitive subjects (Adiels et al. 2006). The hepatic uptake of VLDL, IDL, and LDL is decreased in individuals with metabolic syndrome, this results in increased circulation time of these lipoproteins (Chan et al. 2004) which further adds to the increased plasma levels. The formation of sdLDL involves both cholesteryl ester transfer protein (CETP) and hepatic lipase: CETP facilitates the transfer of triglycerides from VLDL1 to LDL; the resulting triglyceride-rich LDL is a preferred substrate for hepatic lipase; and increased lipolysis of triglyceride-rich LDL results in the formation of sdLDL (Verges 2005). Consequently, it appears that the presence of large triglyceride rich VLDL1 particles is a precondition for sdLDL formation (Adiels et al. 2006). Nevertheless, sdLDL are also observed in patients with type 2 diabetes and insulin resistance with close to normal triglyceride levels. This may be explained by increased hepatic lipase activity (Feingold et al. 1992). It has been shown that the presence of sdLDL particles is associated with increased cardiovascular risk (Vakkilainen et al. 2003), but it is uncertain whether sdLDL levels add independent information on risk assessment over standard risk factors (Sacks et al. 2003). Recently, data from the Coronary Artery Risk

Development in Young Adults (CARDIA), a population-based, prospective, observational study, showed that higher concentrations of oxidized LDL were associated with augmented incidence of metabolic syndrome taken as a whole, as well as its components: abdominal obesity, hyperglycaemia, and hypertriglyceridaemia. It is not yet possible to conclude whether oxidized LDL is a marker linked to the causal factors on the pathway to the development of metabolic syndrome, or whether it is by itself a mediator in this pathway. However, the strong association of oxidized LDL with the incidence of metabolic syndrome seems consistent with a causal role (Holvoet et al. 2008). Circulating oxidized LDL levels are coupled with obesity and weight loss results in a decrease of oxidized LDL (Weinbrenner et al. 2006). This association may be explained by the presence of sdLDL that are more prone to oxidation (Lamarche 1998). An additional possible explanation is that adipose tissue contributes to the oxidation of LDL by 2 biochemical actions: increased adiposity may boost production of arachidonate-5-lipoxygenase, which catalyzes LDL oxidation; increased adiposity may decrease production of superoxide dismutase, which protects LDL against oxidation (Verreth et al. 2004).

3.5.4 Decreased Antiatherogenic Activities of HDL in Metabolic Syndrome

High density lipoproteins exert several antiatherogenic activities like cellular cholesterol efflux, together with antioxidative, anti-inflammatory, and anti-thrombotic actions (Assmann 2003). One of the vasculoprotective properties of HDL particles is based on the potent protection of endothelial cells from the cytotoxic effects of oxLDL (Salvayre et al. 2002). In fact, HDL protect endothelial cells both from apoptosis induced by mildly oxidized LDL (Suc et al. 1997), tumour necrosis factor-α (Sugano et al. 2000) or growth factor deficiency (Nofer et al. 2001) and from necrotic cell death (Kimura et al. 2003). As previously described regarding the formation of sdLDL metabolism, increased levels of VLDL1 also alter the composition of HDL through the actions of CETP and hepatic lipase, leading to the formation of small, dense HDL (sdHDL) and increased catabolism of these particles (Rashid et al. 2003). The replacement of cholesterol esters by triglycerides mediated by CETP decreases the conformational stability of the central and C-terminal domains of apoA-I which are critical for HDL particles to act as lipid acceptors. In particular, triglyceride enrichment reduces exposure of apoA-I to the aqueous phase, because of its further penetration into the lipid core of HDL (Curtiss et al. 2000). The changes in HDL chemical composition are coupled with the diminished antioxidative activity of small HDL3. This occurs in atherogenic low HDL-C dyslipidaemias observed in metabolic syndrome and type 2 diabetes (Hansel et al. 2004). Oxidative stress is associated with covalent modifications of functional amino acid residues in apoA-I resulting in the impairment of HDL capacity to promote cellular cholesterol efflux (Zheng et al. 2004). Paraoxonase-1 (PON1) is a HDL associated enzyme that, among many other functions, metabolizes pro-inflammatory lipids formed during the oxidation of LDL, and is therefore potentially anti-atherogenic (Mackness et al. 1993). However, HDL associated enzymes possessing antioxidative activities, such as PON1 and platelet-activating factor acetylhydrolase, do not appear to contribute to functional deficiency of small HDL in metabolic syndrome. In fact, no decrease in the activities of these enzymes was observed in any HDL subfraction from metabolic syndrome patients vs. controls (Hansel et al. 2004). Metabolic syndrome patients have small dense HDL particles that are deficient in their ability to protect endothelial cells from apoptosis induced by oxidized LDL. Restoring the biological activities of HDL particles is therefore proposed as a new therapeutic approach to reduce cardiovascular risk in this insulin-resistant phenotype (de Souza et al. 2008).

3.5.5 Antioxidants, Dyslipidaemia and Atherosclerosis

Favourable epidemiological evidence and success in several animal trials in a number of species using a variety of antioxidants, made the hypothesis of treating atherosclerosis with antioxidants apparently strong. This initial success led to clinical trials that were ill-conceived and were performed without comprehending the enzymes and factors involved in the oxidative process including those added by associated risk factors; the nature of the oxidative process and associated changes in the lipid and protein moieties; whether other lipoproteins, namely HDL were affected by these processes; whether antioxidants improved any specific steps in the oxidative process; the compatibility of the oxidative process/antioxidant action with current therapy; the steps involved in the atherogenic process and the role(s) of oxidative processes, if any, on these steps; the distinction between the early fatty streak lesions and the advanced vulnerable lesions that lead to plaque rupture, thrombosis, and acute coronary events, and above all the distinction between the animal models and human atherosclerotic development (Parthasarathy et al. 2008).

3.5.6 Effects of Vitamin E on Lipidaemia

The majority of clinical trials on the hypolipidaemic effects of antioxidant vitamin E were not able to correlate supplementation in hyperlipidaemic patients with lower levels of plasma lipids. Supplementation of low-fat fed hypercholesterolemic patients with α-, γ- and δ-tocotrienyl acetates did not lower either total and LDL cholesterol levels, or apo-B levels (Shidfar et al. 2003). No significant changes in total, HDL, LDL cholesterol, and triglycerides levels were observed in patients with coronary heart disease supplemented with vitamin E either (Gupta et al. 2001). Similarly, the large-scale ASAP study showed no effect of vitamin E supplementation for 6 years on HDL cholesterol (Salonen et al. 2003). Nevertheless, some studies have confirmed the hypocholesterolaemic effects of tocotrienol rich fractions and individual tocotrienols in experimental animals (Qureshi et al. 1991a; Watkins et al. 1993) and humans (Qureshi et al. 1991b). The tocotrienol-rich fraction mixture of palm oil containing high-tocopherol concentration had no impact on serum total cholesterol or LDL-cholesterol in hypercholesterolaemic human subjects (Mensink et al. 1999). A rapidly expanding body of evidence supports that members of the vitamin E family are functionally unique. In recognition of this fact, title claims in publications should be limited to the specific form of vitamin E studied (Sen et al. 2007). Many studies on the anti-atherosclerotic effects of vitamin E did not exclude the anti-atherosclerotic and anti-oxidative effects of statins, this could have reduced the possibility that a prespecified sample size had an adequate power to observe a difference between vitamin E and placebo-treated groups (Violi et al. 2008).

3.5.7 Effects of Vitamin C on Lipidaemia

Some studies support vitamin C as a promising lipid lowering factor. In a double blind, placebo trial of parallel design, 68 hyperlipidaemic patients were supplemented with n3-fatty acids alone, vitamin C alone and their combination for 10 weeks. There was a significant difference in the blood vitamin C level at the end of the study compared to the initial value in groups given the antioxidant supplement alone or combined, compared to the control value and that of the n3-fatty acids group. A significant decrease in apo-B and total cholesterol was observed in the vitamin C group at the end of the study compared to the respective initial values, but not in the combination group (Shidfar et al. 2003). In another study, supplementation with high (500 mg/day) and low (50 mg/day) doses of ascorbic acid resulted in no favourable effect on mean serum concentrations of total cholesterol, HDL and LDL-cholesterol, and triglycerides, although high-dose supplementation increased serum vitamin C concentrations substantially. Nonetheless, in the group supplemented with high dose of vitamin C, the mean serum triglyceride levels decreased, while in the hypertriglyceridemic subjects, the mean reduction was statistically significant (Kim et al. 2004).

3.5.8 Effects of Flavonoids on Lipidaemia

As with other antioxidants, results from clinical trials studying the hypolipidaemic activity of flavonoids are controversial. Administration of the hydroalcoholic extract of *Achillea wilhelmsii* C. Koch, a plant full of flavonoids, to 120 male and female moderate hyperlipidaemic and primary hypertensive subjects, resulted in a significant decrease in triglycerides after 2 months and in a significant decrease in total cholesterol and LDL-cholesterol after 4 months. A significant increase in HDL-cholesterol was observed after 6 months of treatment (Asgary et al. 2000). In a double-blind, randomized, placebo controlled, parallel-group trial set with 240 men and women on a low-fat diet with mild to moderate hypercholesterolaemia, daily administration of a capsule containing theaflavin-enriched green tea extract

(375 mg) for 12 weeks resulted in changes in total cholesterol, LDL-cholesterol, HDL-cholesterol, and triglyceride levels compared with baseline (Maron et al. 2003). In normocholesterolemic and mildly hypercholesterolaemic postmenopausal women a high-isoflavone diet resulted in lower LDL cholesterol and lower ratio of LDL to HDL cholesterol. Isoflavone consumption did not significantly affect plasma concentrations of total or HDL cholesterol, triglycerides, apolipoprotein A–I, apolipoprotein B, and lipoprotein(a), respectively (Wangen et al. 2001). Phytoestrogen supplementation (150 mg/day) in moderately hypercholesterolemic, elderly, postmenopausal women, resulted in no significant differences in total triglycerides, total cholesterol or HDL cholesterol after 2 months of treatment or even after treatment for over a 6-month period (Dewell et al. 2002). In moderately hypercholesterolemic subjects, soy-derived isoflavones had no significant effect on total cholesterol, LDL and HDL cholesterol, and triglyceride concentrations (Lichtenstein et al. 2002). Whole soy supplementation in mildly hypercholesterolaemic and/or hypertensive volunteers markedly increased plasma and urinary isoflavones, but there were no differences in plasma lipids (Meyer et al. 2004).

3.5.9 Effects of Carotenoids on Lipidaemia

The antioxidant carotenoid lycopene has shown some hypocholesterolemic effectiveness. In a study with six healthy male subjects supplemented with 60 mg/day of lycopene for 3 months, a significant 14% reduction in their plasma LDL cholesterol was observed (Fuhrman et al. 1997). However, in a randomized cross-over dietary intervention study, supplementation with lycopene from tomato products (20–150 mg/day) of healthy subjects failed to modify serum total cholesterol and LDL and HDL cholesterols (Agarwal et al. 1998).

All the above mentioned data strongly support the involvement of oxidative stress in many features of the metabolic syndrome; however, there is still too much controversy to unequivocally support antioxidant supplementation strategies to prevent or treat this syndrome.

3.6 Concluding Remarks

Obesity is the most common disorder in both developed and, also, developing countries and is associated by a reduction in insulin sensitivity. Furthermore, the degree of visceral adiposity conveys an independent prediction of risk beyond body mass index (BMI) for cardiovascular disease.

The molecular mechanisms involved in obesity-related insulin resistance are not well understood. However, it has been well demonstrated that adipocytes are able to synthesize and secrete several cytokines (adipokines), such as leptin, TNFα and IL-6 (Vincent and Taylor 2006) and the hypothesis proposing that these adipokines may be responsible for insulin resistance in obesity (Bastard et al. 2006), in an oxidative stress-dependent way, has emerged.

Taken together, the existence of a link among the hyperglycaemia- and FFA-induced increases in ROS and oxidative stress, activation of stress-sensitive pathways, and the eventual development of not only the late complications of diabetes, but also insulin resistance and β-cell dysfunction seem highly plausible.

Although the understanding about mechanisms of how hyperglycaemia induces oxidative stress and, consequently, pathology has advanced considerably in last years, effective therapeutic strategies to prevent or delay damage remain limited. Additional research is urgently needed.

High blood pressure (BP) is one of the main metabolic syndrome components, and metabolic syndrome has been found in about 30–40% of hypertensives. There is still a strong debate as to whether or not the presence of metabolic syndrome increases the hypertension-induced cardiovascular risk (Kahn et al. 2005).

It also remains unclear whether elevated levels of free radicals initiate the development of hypertension, are a consequence of the disease process itself or both (Grossman 2008).

However, there is strong evidence that oxidative stress may add to the generation and/or maintenance of hypertension via several possible mechanisms (McIntyre et al. 1999).

The data from many clinical studies supporting antioxidant supplementation to lower blood pressure is limited. This data scarcity hampers the possibility of establishing antioxidant therapies as a useful tool to treat hypertension. Some authors even consider hypertension as the cause of oxidative stress instead of a consequence (Grossman 2008).

Dyslipidaemia is probably the major mediator of atherogenicity in the metabolic syndrome (Alexander et al. 2003).

The LDL oxidation hypothesis formulated in the 1980's to explain the formation of fatty streak lesions is now supported by thousands of publications providing evidence for the presence of oxidative processes in atherosclerotic human disease (Parthasarathy et al. 2008; Rosin 2007). As it happened with hypertension, antioxidant supplementation to prevent or treat atherosclerosis produced mixed results. There is still too much controversy to unequivocally support antioxidant supplementation strategies to prevent or treat hypertension and dyslipidaemia, either as isolated features or as components of the metabolic syndrome.

References

Abdilla N, Tormo MC, Fabia MJ et al. *Impact of the components of metabolic syndrome on oxidative stress and enzymatic antioxidant activity in essential hypertension.* J Hum Hypertens. 2007; 21: 68–75.

Adiels M, Boren J, Caslake MJ et al. *Overproduction of VLDL1 driven by hyperglycaemia is a dominant feature of diabetic dyslipidaemia.* Arterioscler Thromb Vasc Biol 2005; 25: 1697–703.

Adiels M, Olofsson SO, Taskinen MR et al. *Diabetic dyslipidaemia.* Curr Opin Lipidol. 2006; 17: 238–46.

Adiels M, Olofsson SO, Taskinen MR et al. *Overproduction of very low-density lipoproteins is the hallmark of the dyslipidaemia in the metabolic syndrome.* Arterioscler Thromb Vasc Biol. 2008; 28: 1225–36.

Adiels M, Taskinen MR, Packard C et al. Overproduction of large VLDL particles is driven by increased liver fat content in man. Diabetologia. 2006; 49: 755–65.

Agarwal S, Rao AV. *Tomato lycopene and low density lipoprotein oxidation: a human dietary intervention study.* Lipids. 1998; 33: 981–4.

Alexander CM, Landsman PB, Teutsch SM et al. *NCEP-defined metabolic syndrome, diabetes, and prevalence of coronary heart disease among NHANES III participants age 50 years and older.* Diabetes. 2003; 52: 1210–4.

Andrew PJ, Mayer B. *Enzymatic function of nitric oxide synthases.* Cardiovasc Res 1999; 43: 521–31.

Asgary S, Naderi GH, Sarrafzadegan N et al. *Antihypertensive and antihyperlipidaemic effects of Achillea wilhelmsii.* Drugs Exp Clin Res. 2000; 26: 89–93.

Assmann G, Nofer JR. *Atheroprotective effects of high-density lipoproteins.* Annu Rev Med. 2003; 54: 321–41.

Avogaro P, Bon GB, Cazzolato G. *Presence of a modified low density lipoprotein in humans.* Arteriosclerosis. 1988; 8: 79–87.

Bastard JP, Maachi M, Lagathu C, Kim MJ, Caron M, Vidal H, Capeau J, Feve B.*Recent advances in the relationship between obesity, inflammation, and insulin resistance.* Eur Cytokine Netw. 2006; 17: 4–12.

Baynes JW, Thorpe SR. *Role of oxidative stress in diabetic complications: a new perspective on an old paradigm.* Diabetes. 1999; 48: 1–9.

Bayraktutan U, Blayney L, Shah AM. *Molecular characterization and localization of the NAD(P)H oxidase components gp91-phox and p22-phox in endothelial cells.* Arterioscler Thromb Vasc Biol. 2000; 20: 1903–11.

Bays HE, González-Campoy JM, Bray GA, Kitabchi AE, Bergman DA, Schorr AB, Rodbard HW, Henry RR. *Pathogenic potential of adipose tissue and metabolic consequences of adipocyte hypertrophy and increased visceral adiposity.* Expert Rev Cardiovasc Ther. 2008; 6: 343–68.

Beckman JS, Koppenol WH. *Nitric oxide, superoxide, and peroxynitrite: the good, the bad, and ugly.* Am J Physiol. 1996; 271: C1424–37.

Bełtowski J, Wójcicka G, Górny D, Marciniak A. *The effect of dietary-induced obesity on lipid peroxidation, antioxidant enzymes and total plasma antioxidant capacity.* J Physiol Pharmacol. 2000; 51: 883–96.

Boo YC, Jo H. *Flow-dependent regulation of endothelial nitric oxide synthase: role of protein kinases.* Am J Physiol Cell Physiol. 2003; 285: C499–508.

Borcea V, Nourooz-Zadeh J, Wolff SP, Klevesath M, Hofmann M, Urich H, Wahl P, Ziegler R, Tritschler H, Halliwell B, Nawroth PP. *Alpha-Lipoic acid decreases oxidative stress even in diabetic patients with poor glycemic control and albuminuria.* Free Radic Biol Med. 1999; 26: 1495–500.

Bougoulia M, Triantos A, Koliakos G. *Plasma interleukin-6 levels, glutathione peroxidase and isoprostane in obese women before and after weight loss. Association with cardiovascular risk factors.* Hormones. 2006; 5: 192–199.

Cai H, McNally JS, Weber M et al. *Oscillatory shear stress upregulation of endothelial nitric oxide synthase requires intracellular hydrogen peroxide and CaMKII.* J Mol Cell Cardiol. 2004; 37: 121–5.

Cameron NE, Cotter MA. *Metabolic and vascular factors in the pathogenesis of diabetic neuropathy.* Diabetes. 1997; 46 Suppl 2: S31–7.

Cardona F, Túnez I, Tasset I, Montilla P, Collantes E, Tinahones FJ. *Fat overload aggravates oxidative stress in patients with the metabolic syndrome.* Eur J Clin Invest. 2008b; 38: 510–5.

Cardona F, Túnez I, Tasset I, Murri M, Tinahones FJ. *Similar increase in oxidative stress after fat overload in persons with baseline hypertriglyceridemia with or without the metabolic syndrome.* Clin Biochem. 2008a; 41: 701–5.

Chan CB, Harper ME. *Uncoupling proteins: role in insulin resistance and insulin insufficiency.* Curr Diabetes Rev. 2006; 2: 271–83.

Chan DC, Barrett PH, Watts GF. *Lipoprotein kinetics in the metabolic syndrome: pathophysiological and therapeutic lessons from stable isotope studies.* Clin Biochem Rev. 2004; 25: 31–48.

Chen X, Touyz RM, Park JB et al. *Antioxidant effects of vitamins C and E are associated with altered activation of vascular NADPH oxidase and superoxide dismutase in stroke-prone SHR.* Hypertension. 2001; 38: 606–11.

Cracowski JL, Durand T, Bessard G. *Isoprostanes as a biomarker of lipid peroxidation in humans: physiology, pharmacology and clinical implications.* Trends Pharmacol Sci. 2002; 23: 360–6.

Croft KD, McGiff JC, Sanchez-Mendoza A et al. *Angiotensin II releases 20-HETE from rat renal microvessels.* Am J Physiol Renal Physiol. 2000; 279: F544–51.

Curtiss LK, Bonnet DJ, Rye KA. *The conformation of apolipoprotein A-I in high-density lipoproteins is influenced by core lipid composition and particle size: a surface plasmon resonance study.* Biochemistry. 2000; 39: 5712–21.

Cuspidi C, Sala C, Zanchetti A. *Metabolic syndrome and target organ damage: role of blood pressure.* Expert Rev Cardiovasc Ther. 2008; 6: 731–43.

Czernichow S, Bertrais S, Blacher J et al. *Effect of supplementation with antioxidants upon long-term risk of hypertension in the SU.VI.MAX study: association with plasma antioxidant levels.* J Hypertens. 2005; 23: 2013–8.

Dandona P, Kumar V, Aljada A et al. *Angiotensin II receptor blocker valsartan suppresses reactive oxygen species generation in leukocytes, nuclear factor-kappa B, in mononuclear cells of normal subjects: evidence of an antiinflammatory action.* J Clin Endocrinol Metab. 2003; 88: 4496–501.

Dandona P, Mohanty P, Ghanim H et al. *The suppressive effect of dietary restriction and weight loss in the obese on the generation of reactive oxygen species by leukocytes, lipid peroxidation, and protein carbonylation.* J Clin Endocrinol Metab. 2001; 86: 355–62.

Davì G, Ciabattoni G, Consoli A, Mezzetti A, Falco A, Santarone S, Pennese E, Vitacolonna E, Bucciarelli T, Costantini F, Capani F, Patrono C.*In vivo formation of 8-iso-prostaglandin f2alpha and platelet activation in diabetes mellitus: effects of improved metabolic control and vitamin E supplementation.* Circulation. 1999; 99: 224–9.

de Ferranti S, Mozaffarian D. *The perfect storm: obesity, adipocyte dysfunction, and metabolic consequences.* Clin Chem. 2008; 54: 945–55.

de Souza JA, Vindis C, Hansel B et al. *Metabolic syndrome features small, apolipoprotein A-I-poor, triglyceride-rich HDL3 particles with defective anti-apoptotic activity.* Atherosclerosis. 2008; 197: 84–94.

Dewell A, Hollenbeck CB, Bruce B. *The effects of soy-derived phytoestrogens on serum lipids and lipoproteins in moderately hypercholesterolemic postmenopausal women.* J Clin Endocrinol Metab. 2002; 87: 118–21.

Engeli S, Negrel R, Sharma AM. *Physiology and pathophysiology of the adipose tissue renin-angiotensin system.* Hypertension. 2000; 35: 1270–7.

Evans JL, Goldfine ID, Maddux BA, Grodsky GM.*Are oxidative stress-activated signaling pathways mediators of insulin resistance and beta-cell dysfunction?* Diabetes. 2003; 52: 1–8.

Evans JL, Goldfine ID, Maddux BA, Grodsky GM. *Oxidative stress and stress-activated signaling pathways: a unifying hypothesis of type 2 diabetes.* Endocr Rev. 2002; 23(5):599–622.

Evans JL, Goldfine ID. *Alpha-lipoic acid: a multifunctional antioxidant that improves insulin sensitivity in patients with type 2 diabetes.* Diabetes Technol Ther. 2000; 2: 401–13.

Evans JL, Maddux BA, Goldfine ID. *The molecular basis for oxidative stress-induced insulin resistance.* Antioxid Redox Signal. 2005; 7: 1040–52.

Ezzati M, Lopez AD, Rodgers A et al. *Selected major risk factors and global and regional burden of disease.* Lancet. 2002; 360: 1347–60.

Feingold KR, Grunfeld C, Pang M et al. *LDL subclass phenotypes and triglyceride metabolism in non-insulin-dependent diabetes.* Arterioscler Thromb. 1992; 12: 1496–502.

Fleming I, Michaelis UR, Bredenkotter D et al. *Endothelium-derived hyperpolarizing factor synthase (Cytochrome P450 2C9) is a functionally significant source of reactive oxygen species in coronary arteries.* Circ Res. 2001; 88: 44–51.

Ford ES. *Intake and circulating concentrations of antioxidants in metabolic syndrome.* Curr Atheroscler Rep. 2006; 8: 448–52.

Fridlyand LE, Philipson LH. *Reactive species, cellular repair and risk factors in the onset of type 2 diabetes mellitus: review and hypothesis.* Curr Diabetes Rev. 2006; 2: 241–59.

Fuhrman B, Elis A, Aviram M. *Hypocholesterolemic effect of lycopene and beta-carotene is related to suppression of cholesterol synthesis and augmentation of LDL receptor activity in macrophages.* Biochem Biophys Res Commun. 1997; 233: 658–62.

Fukui T, Ishizaka N, Rajagopalan S et al. *p22phox mRNA expression and NADPH oxidase activity are increased in aortas from hypertensive rats.* Circ Res 1997; 80: 45–51.

Furukawa S, Fujita T, Shimabukuro M, Iwaki M, Yamada Y, Nakajima Y, Nakayama O, Makishima M, Matsuda M, Shimomura I. *Increased oxidative stress in obesity and its impact on metabolic syndrome.* J Clin Invest. 2004; 114: 1752–61.

Ginsberg HN, Zhang YL, Hernandez-Ono A. *Metabolic syndrome: focus on dyslipidaemia.* Obesity (Silver Spring) 2006; 14 Suppl 1:41S–9S.

Gregor MF, Hotamisligil GS. *Thematic review series: Adipocyte Biology. Adipocyte stress: the endoplasmic reticulum and metabolic disease.* J Lipid Res. 2007; 48: 1905–14.

Grossman E. *Does increased oxidative stress cause hypertension?* Diabetes Care. 2008; 31 Suppl 2:S185–9.

Grundy SM. *Hypertriglyceridemia, atherogenic dyslipidaemia, and the metabolic syndrome.* Am J Cardiol. 1998; 81: 18B–25B.

Gupta R, Singhal S, Goyle A et al. *Antioxidant and hypocholesterolaemic effects of Terminalia arjuna tree-bark powder: a randomised placebo-controlled trial.* J Assoc Physicians India. 2001; 49: 231–5.

Haberland ME, Fogelman AM, Edwards PA. *Specificity of receptor-mediated recognition of malondialdehyde-modified low density lipoproteins.* Proc Natl Acad Sci USA. 1982; 79: 1712–6.

Haberland ME, Olch CL, Folgelman AM. *Role of lysines in mediating interaction of modified low density lipoproteins with the scavenger receptor of human monocyte macrophages.* J Biol Chem. 1984; 259: 11305–11.

Hansel B, Giral P, Nobecourt E, Chantepie S, Bruckert E, Chapman MJ, Kontush A. *Metabolic syndrome is associated with elevated oxidative stress and dysfunctional dense high-density lipoprotein particles displaying impaired antioxidative activity.* J Clin Endocrinol Metab. 2004; 89: 4963–71.

Harrison DG, Widder J, Grumbach I et al. *Endothelial mechanotransduction, nitric oxide and vascular inflammation.* J Intern Med 2006; 259: 351–63.

Harte A, McTernan P, Chetty R et al. *Insulin-mediated upregulation of the renin angiotensin system in human subcutaneous adipocytes is reduced by rosiglitazone.* Circulation 2005; 111: 1954–61.

Henriksen T, Mahoney EM, Steinberg D. *Enhanced macrophage degradation of biologically modified low density lipoprotein.* Arteriosclerosis. 1983; 3: 149–59.

Holvoet P, Lee DH, Steffes M et al. *Association between circulating oxidized low-density lipoprotein and incidence of the metabolic syndrome.* Jama. 2008; 299: 2287–93.

Holvoet P, Vanhaecke J, Janssens S et al. *Oxidized LDL and malondialdehyde-modified LDL in patients with acute coronary syndromes and stable coronary artery disease.* Circulation. 1998; 98: 1487–94.

Hong HJ, Hsiao G, Cheng TH et al. *Supplemention with tetrahydrobiopterin suppresses the development of hypertension in spontaneously hypertensive rats.* Hypertension. 2001; 38: 1044–8.

Hotta N. *New approaches for treatment in diabetes: aldose reductase inhibitors.* Biomed Pharmacother. 1995; 49: 232–43.

Ido Y, Kilo C, Williamson JR. *Cytosolic NADH/NAD+, free radicals, and vascular dysfunction in early diabetes mellitus.* Diabetologia. 1997; 40 Suppl 2: S115–7.

Inoue N, Ramasamy S, Fukai T et al. *Shear stress modulates expression of Cu/Zn superoxide dismutase in human aortic endothelial cells.* Circ Res. 1996; 79: 32–7.

Ishii H, Koya D, King GL. *Protein kinase C activation and its role in the development of vascular complications in diabetes mellitus.* J Mol Med. 1998; 76: 21–31.

Jung O, Schreiber JG, Geiger H et al. *gp91phox-containing NADPH oxidase mediates endothelial dysfunction in renovascular hypertension.* Circulation. 2004; 109: 1795–801.

Kahn R, Buse J, Ferrannini E et al. *The metabolic syndrome: time for a critical appraisal. Joint statement from the American Diabetes Association and the European Association for the Study of Diabetes.* Diabetologia. 2005; 48: 1684–99.

Kedziora-Kornatowska K, Czuczejko J, Pawluk H et al. *The markers of oxidative stress and activity of the antioxidant system in the blood of elderly patients with essential arterial hypertension.* Cell Mol Biol Lett. 2004; 9: 635–41.

Khan BV, Navalkar S, Khan QA et al. *Irbesartan, an angiotensin type 1 receptor inhibitor, regulates the vascular oxidative state in patients with coronary artery disease.* J Am Coll Cardiol. 2001; 38: 1662–7.

Kim JA, Wei Y, Sowers JR. *Role of mitochondrial dysfunction in insulin resistance.* Circ Res. 2008; 102: 401–14.

Kim MK, Sasaki S, Sasazuki S et al. *Long-term vitamin C supplementation has no markedly favourable effect on serum lipids in middle-aged Japanese subjects.* Br J Nutr. 2004; 91: 81–90.

Kimura T, Sato K, Malchinkhuu E et al. *High-density lipoprotein stimulates endothelial cell migration and survival through sphingosine 1-phosphate and its receptors.* Arterioscler Thromb Vasc Biol. 2003; 23: 1283–8.

Kojda G, Hambrecht R. *Molecular mechanisms of vascular adaptations to exercise. Physical activity as an effective antioxidant therapy?* Cardiovasc Res. 2005; 67: 187–97.

Laakso J, Mervaala E, Himberg JJ et al. *Increased kidney xanthine oxidoreductase activity in salt-induced experimental hypertension.* Hypertension. 1998; 32: 902–6.

Lacy F, O'Connor DT, Schmid-Schonbein GW. *Plasma hydrogen peroxide production in hypertensives and normotensive subjects at genetic risk of hypertension.* J Hypertens. 1998; 16: 291–303.

Lamarche B. *Abdominal obesity and its metabolic complications: implications for the risk of ischaemic heart disease.* Coron Artery Dis. 1998; 9: 473–81.

Landmesser U, Cai H, Dikalov S et al. *Role of p47(phox) in vascular oxidative stress and hypertension caused by angiotensin II.* Hypertension. 2002; 40: 511–5.

Landmesser U, Dikalov S, Price SR et al. *Oxidation of tetrahydrobiopterin leads to uncoupling of endothelial cell nitric oxide synthase in hypertension.* J Clin Invest. 2003; 111: 1201–9.

Lassegue B, Clempus RE. *Vascular NAD(P)H oxidases: specific features, expression, and regulation.* Am J Physiol Regul Integr Comp Physiol. 2003; 285: R277–97.

Laursen JB, Somers M, Kurz S et al. *Endothelial regulation of vasomotion in apoE-deficient mice: implications for interactions between peroxynitrite and tetrahydrobiopterin.* Circulation. 2001; 103: 1282–8.

Lawes CM, Vander Hoorn S, Rodgers A. *Global burden of blood-pressure-related disease, 2001.* Lancet. 2008; 371: 1513–8.

Leclerc PC, Proulx CD, Arguin G et al. *Ascorbic acid decreases the binding affinity of the AT1 receptor for angiotensin II.* Am J Hypertens. 2008; 21: 67–71.

Lee DH, Gross MD, Jacobs DR Jr. *Association of serum carotenoids and tocopherols with gamma-glutamyltransferase: the Cardiovascular Risk Development in Young Adults (CARDIA) Study.* Clin Chem. 2004; 50: 582–8.

Li JM, Shah AM. *Intracellular localization and preassembly of the NADPH oxidase complex in cultured endothelial cells.* J Biol Chem. 2002; 277: 19952–60.

Lichtenstein AH, Jalbert SM, Adlercreutz H et al. *Lipoprotein response to diets high in soy or animal protein with and without isoflavones in moderately hypercholesterolemic subjects.* Arterioscler Thromb Vasc Biol. 2002; 22: 1852–8.

Lyons TJ, Jenkins AJ. *Glycation, oxidation, and lipoxidation in the development of the complications of diabetes: a carbonyl stress hypothesis.* Diabetes Rev. 1997; 5: 365–391.

Maassen JA, Romijn JA, Heine RJ. *Fatty acid-induced mitochondrial uncoupling in adipocytes as a key protective factor against insulin resistance and beta cell dysfunction: a new concept in the pathogenesis of obesity-associated type 2 diabetes mellitus.* Diabetologia. 2007; 50: 2036–41.

Mackness MI, Arrol S, Abbott C et al. *Protection of low-density lipoprotein against oxidative modification by high-density lipoprotein associated paraoxonase.* Atherosclerosis. 1993; 104: 129–35.

Maron DJ, Lu GP, Cai NS et al. *Cholesterol-lowering effect of a theaflavin-enriched green tea extract: a randomized controlled trial.* Arch Intern Med. 2003; 163: 1448–53.

Martyn JA, Kaneki M, Yasuhara S. *Obesity-induced insulin resistance and hyperglycaemia: etiologic factors and molecular mechanisms.* Anesthesiology. 2008; 109: 137–48.

McIntyre M, Bohr DF, Dominiczak AF.*Endothelial function in hypertension: the role of superoxide anion.* Hypertension. 1999; 34 (4 Pt 1): 539–45.

Mensink RP, van Houwelingen AC, Kromhout D et al. *A vitamin E concentrate rich in tocotrienols had no effect on serum lipids, lipoproteins, or platelet function in men with mildly elevated serum lipid concentrations.* Am J Clin Nutr. 1999; 69: 213–9.

Mervaala EM, Cheng ZJ, Tikkanen I et al. *Endothelial dysfunction and xanthine oxidoreductase activity in rats with human renin and angiotensinogen genes.* Hypertension. 2001; 37 (2 Part 2): 414–8.

Meydani M. *Vitamin E modulation of cardiovascular disease.* Ann N Y Acad Sci. 2004; 1031:271–9.

Meyer BJ, Larkin TA, Owen AJ et al. *Limited lipid-lowering effects of regular consumption of whole soybean foods.* Ann Nutr Metab. 2004; 48: 67–78.

Miller ER, 3rd, Pastor-Barriuso R, Dalal D et al. *Meta-analysis: high-dosage vitamin E supplementation may increase all-cause mortality.* Ann Intern Med. 2005; 142 (1):37–46.

Minuz P, Patrignani P, Gaino S et al. *Increased oxidative stress and platelet activation in patients with hypertension and renovascular disease.* Circulation. 2002; 106(22): 2800–5.

Mohanty P, Ghanim H, Hamouda W et al. *Both lipid and protein intakes stimulate increased generation of reactive oxygen species by polymorphonuclear leukocytes and mononuclear cells.* Am J Clin Nutr. 2002; 75: 767–72.

Heart Protection Study Collaborative Group. *MRC/BHF Heart Protection Study of antioxidant vitamin supplementation in 20,536 high-risk individuals: a randomised placebo-controlled trial.* Lancet. 2002; 360: 23–33.

Nakazono K, Watanabe N, Matsuno K et al. *Does superoxide underlie the pathogenesis of hypertension?* Proc Natl Acad Sci USA. 1991; 88: 10045–8.

Nofer JR, Levkau B, Wolinska I et al. *Suppression of endothelial cell apoptosis by high density lipoproteins (HDL) and HDL-associated lysosphingolipids.* J Biol Chem. 2001; 276(37): 34480–5.

Nourooz-Zadeh J, Tajaddini-Sarmadi J, McCarthy S, Betteridge DJ, Wolff SP. *Elevated levels of authentic plasma hydroperoxides in NIDDM.* Diabetes. 1995; 44: 1054–8.

Orie NN, Zidek W, Tepel M. *Reactive oxygen species in essential hypertension and non-insulin-dependent diabetes mellitus.* Am J Hypertens. 1999; 12(12 Pt 1-2): 1169–74.

Ortiz MC, Manriquez MC, Romero JC et al. *Antioxidants block angiotensin II-induced increases in blood pressure and endothelin.* Hypertension. 2001; 38 (3 Pt 2):655–9.

Paolisso G, Giugliano D. *Oxidative stress and insulin action: is there a relationship?* Diabetologia. 1996; 39: 357–63.

Parthasarathy S, Litvinov D, Selvarajan K et al. *Lipid peroxidation and decomposition–conflicting roles in plaque vulnerability and stability.* Biochim Biophys Acta. 2008; 1781 (5): 221–31.

Pedro-Botet J, Covas MI, Martin S et al. Decreased endogenous antioxidant enzymatic status in essential hypertension. J Hum Hypertens. 2000; 14 (6):343–5.

Pfeiffer A, Schatz H. *Diabetic microvascular complications and growth factors.* Exp Clin Endocrinol Diabetes. 1995; 103: 7–14.

Plantinga Y, Ghiadoni L, Magagna A et al.*Supplementation with vitamins C and E improves arterial stiffness and endothelial function in essential hypertensive patients.* Am J Hypertens. 2007; 20: 392–7.

Qureshi AA, Qureshi N, Hasler-Rapacz JO et al. *Dietary tocotrienols reduce concentrations of plasma cholesterol, apolipoprotein B, thromboxane B2, and platelet factor 4 in pigs with inherited hyperlipidaemias.* Am J Clin Nutr. 1991a; 53 (4 Suppl):1042S–6S.

Qureshi AA, Qureshi N, Wright JJ et al. *Lowering of serum cholesterol in hypercholesterolemic humans by tocotrienols (palmvitee).* Am J Clin Nutr. 1991b; 53(4 Suppl): 1021S–6S.

Rajagopalan S, Kurz S, Munzel T et al. *Angiotensin II-mediated hypertension in the rat increases vascular superoxide production via membrane NADH/NADPH oxidase activation. Contribution to alterations of vasomotor tone.* J Clin Invest. 1996; 97: 1916–23.

Randle PJ, Kerbey AL, Espinal J. *Mechanisms decreasing glucose oxidation in diabetes and starvation: role of lipid fuels and hormones.* Diabetes Metab Rev. 1988; 4: 623–38.

Rashid S, Watanabe T, Sakaue T et al. *Mechanisms of HDL lowering in insulin resistant, hypertriglyceridemic states: the combined effect of HDL triglyceride enrichment and elevated hepatic lipase activity.* Clin Biochem. 2003; 36: 421–9.

Rey FE, Cifuentes ME, Kiarash A et al. *Novel competitive inhibitor of NAD(P)H oxidase assembly attenuates vascular O(2)(-) and systolic blood pressure in mice.* Circ Res. 2001; 89: 408–14.

Robertson R, Zhou H, Zhang T, Harmon JS. *Chronic oxidative stress as a mechanism for glucose toxicity of the beta cell in type 2 diabetes.* Cell Biochem Biophys. 2007; 48: 139–46.

Rodrigo R, Passalacqua W, Araya J et al. *Implications of oxidative stress and homocysteine in the pathophysiology of essential hypertension.* J Cardiovasc Pharmacol. 2003; 42: 453–61.

Rodrigo R, Prat H, Passalacqua W et al. *Decrease in oxidative stress through supplementation of vitamins C and E is associated with a reduction in blood pressure in patients with essential hypertension.* Clin Sci (Lond). 2008; 114 (10):625–34.

Rodrigo R, Prat H, Passalacqua W et al. *Relationship between oxidative stress and essential hypertension.* Hypertens Res. 2007; 30 (12):1159–67.

Romero JC, Reckelhoff JF. *State-of-the-Art lecture. Role of angiotensin and oxidative stress in essential hypertension.* Hypertension. 1999; 34 (4 Pt 2):943–9.

Rosin BL. *The progression of cardiovascular risk to cardiovascular disease.* Rev Cardiovasc Med. 2007; 8 Suppl 4:S3–8.

Sacks FM, Campos H. *Clinical review 163: Cardiovascular endocrinology: Low-density lipoprotein size and cardiovascular disease: a reappraisal.* J Clin Endocrinol Metab. 2003; 88 (10):4525–32.

Saez GT, Tormos C, Giner V et al. *Factors related to the impact of antihypertensive treatment in antioxidant activities and oxidative stress by-products in human hypertension.* Am J Hypertens. 2004; 17 (9):809–16.

Salonen RM, Nyyssonen K, Kaikkonen J et al. *Six-year effect of combined vitamin C and E supplementation on atherosclerotic progression: the Antioxidant Supplementation in Atherosclerosis Prevention (ASAP) Study.* Circulation. 2003; 107 (7):947–53.

Salvayre R, Auge N, Benoist H et al. *Oxidized low-density lipoprotein-induced apoptosis.* Biochim Biophys Acta. 2002; 1585: 213–21.

Sarkis A, Roman RJ. *Role of cytochrome P450 metabolites of arachidonic acid in hypertension.* Curr Drug Metab. 2004; 5: 245–56.

Sen CK, Khanna S, Rink C et al. *Tocotrienols: the emerging face of natural vitamin E.* Vitam Horm. 2007; 76:203–61.

Seshiah PN, Weber DS, Rocic P et al. *Angiotensin II stimulation of NAD(P)H oxidase activity: upstream mediators.* Circ Res. 2002; 91: 406–13.

Shao B, Oda MN, Vaisar T et al. Pathways for oxidation of high-density lipoprotein in human cardiovascular disease. Curr Opin Mol Ther. 2006; 8: 198–205.

Sharma K, Ziyadeh FN. *Biochemical events and cytokine interactions linking glucose metabolism to the development of diabetic nephropathy.* Semin Nephrol. 1997; 17: 80–92.

Shidfar F, Keshavarz A, Jallali M et al. *Comparison of the effects of simultaneous administration of vitamin C and omega-3 fatty acids on lipoproteins, apo A-I, apo B, and malondialdehyde in hyperlipidaemic patients.* Int J Vitam Nutr Res. 2003; 73: 163–70.

Shoji T, Nishizawa Y, Fukumoto M et al. *Inverse relationship between circulating oxidized low density lipoprotein (oxLDL) and anti-oxLDL antibody levels in healthy subjects.* Atherosclerosis. 2000; 148: 171–7.

Simic DV, Mimic-Oka J, Pljesa-Ercegovac M et al. *Byproducts of oxidative protein damage and antioxidant enzyme activities in plasma of patients with different degrees of essential hypertension.* J Hum Hypertens. 2006; 20: 149–55.

Sjogren P, Basu S, Rosell M, Silveira A, de Faire U, Vessby B, Hamsten A, Hellenius ML, Fisher RM. *Measures of oxidized low-density lipoprotein and oxidative stress are not related and not elevated in otherwise healthy men with the metabolic syndrome.* Arterioscler Thromb Vasc Biol. 2005; 25: 2580–6.

Skalicky J, Muzakova V, Kandar R, Meloun M, Rousar T, Palicka V. *Evaluation of oxidative stress and inflammation in obese adults with metabolic syndrome.* Clin Chem Lab Med. 2008; 46: 499–505.

Slatter DA, Bolton CH, Bailey AJ.*The importance of lipid-derived malondialdehyde in diabetes mellitus.* Diabetologia. 2000; 43: 550–7.

Steinhubl SR. *Why have antioxidants failed in clinical trials?* Am J Cardiol. 2008; 101: 14D–19D.

Stephens JW, Khanolkar MP, Bain SC. *The biological relevance and measurement of plasma markers of oxidative stress in diabetes and cardiovascular disease.* Atherosclerosis. 2008 (*in press*).

Suc I, Escargueil-Blanc I, Troly M et al. *HDL and ApoA prevent cell death of endothelial cells induced by oxidized LDL.* Arterioscler Thromb Vasc Biol. 1997; 17: 2158–66.

Sugano M, Tsuchida K, Makino N. *High-density lipoproteins protect endothelial cells from tumor necrosis factor-alpha-induced apoptosis.* Biochem Biophys Res Commun. 2000; 272: 872–6.

Sun CW, Alonso-Galicia M, Taheri MR et al. Nitric oxide-20-hydroxyeicosatetraenoic acid interaction in the regulation of K+ channel activity and vascular tone in renal arterioles. Circ Res. 1998; 83: 1069–79.

Suzuki H, DeLano FA, Parks DA et al. *Xanthine oxidase activity associated with arterial blood pressure in spontaneously hypertensive rats.* Proc Natl Acad Sci USA. 1998; 95: 4754–9.

Tiedge M, Lortz S, Drinkgern J, Lenzen S. *Relation between antioxidant enzyme gene expression and antioxidative defense status of insulin-producing cells.* Diabetes. 1997; 46: 1733–42.

Touyz RM. *Reactive oxygen species, vascular oxidative stress, and redox signaling in hypertension: what is the clinical significance?* Hypertension. 2004; 44: 248–52.

Urakawa H, Katsuki A, Sumida Y, Gabazza EC, Murashima S, Morioka K, Maruyama N, Kitagawa N, Tanaka T, Hori Y, Nakatani K, Yano Y, Adachi Y. *Oxidative stress is associated with adiposity and insulin resistance in men.* J Clin Endocrinol Metab. 2003; 88: 4673–6.

Uzun H, Karter Y, Aydin S et al. *Oxidative stress in white coat hypertension; role of paraoxonase.* J Hum Hypertens. 2004; 18: 523–8.

Vakkilainen J, Steiner G, Ansquer JC et al. *Relationships between low-density lipoprotein particle size, plasma lipoproteins, and progression of coronary artery disease: the Diabetes Atherosclerosis Intervention Study (DAIS).* Circulation. 2003; 107: 1733–7.

Van Berkel TJ, De Rijke YB, Kruijt JK. *Different fate in vivo of oxidatively modified low density lipoprotein and acetylated low density lipoprotein in rats. Recognition by various scavenger receptors on Kupffer and endothelial liver cells.* J Biol Chem. 1991; 266: 2282–9.

Vasquez-Vivar J, Kalyanaraman B, Martasek P et al. *Superoxide generation by endothelial nitric oxide synthase: the influence of cofactors.* Proc Natl Acad Sci USA. 1998; 95: 9220–5.

Vaziri ND, Ni Z, Oveisi F et al. *Effect of antioxidant therapy on blood pressure and NO synthase expression in hypertensive rats.* Hypertension. 2000; 36 (6):957–64.

Verges B. *Diabetic dyslipidaemia: insights for optimizing patient management.* Curr Med Res Opin. 2005; 21 Suppl 1: S29–40.

Verreth W, De Keyzer D, Pelat M et al. *Weight-loss-associated induction of peroxisome proliferator-activated receptor-alpha and peroxisome proliferator-activated receptor-gamma correlate with reduced atherosclerosis and improved cardiovascular function in obese insulin-resistant mice.* Circulation. 2004; 110: 3259–69.

Vincent HK, Taylor AG.*Biomarkers and potential mechanisms of obesity-induced oxidant stress in humans.* Int J Obes (Lond). 2006; 30: 400–18.

Violi F, Cangemi R. Statin treatment as a confounding factor in human trials with vitamin E. J Nutr. 2008; 138: 1179–81.

Vlassara H. *Recent progress in advanced glycation end products and diabetic complications.* Diabetes. 1997; 46 Suppl 2: S19–25.

Wangen KE, Duncan AM, Xu X et al. *Soy isoflavones improve plasma lipids in normocholesterolemic and mildly hypercholesterolemic postmenopausal women.* Am J Clin Nutr. 2001; 73: 225–31.

Ward NC, Puddey IB, Hodgson JM et al. *Urinary 20-hydroxyeicosatetraenoic acid excretion is associated with oxidative stress in hypertensive subjects.* Free Radic Biol Med. 2005; 38: 1032–6.

Ward NC, Rivera J, Hodgson J et al. *Urinary 20-hydroxyeicosatetraenoic acid is associated with endothelial dysfunction in humans.* Circulation. 2004; 110: 438–43.

Ward NC, Wu JH, Clarke MW et al. *The effect of vitamin E on blood pressure in individuals with type 2 diabetes: a randomized, double-blind, placebo-controlled trial.* J Hypertens. 2007; 25: 227–34.

Wassmann S, Laufs U, Muller K et al. *Cellular antioxidant effects of atorvastatin in vitro and in vivo.* Arterioscler Thromb Vasc Biol. 2002; 22: 300–5.

Watkins T, Lenz P, Gapor A et al. *gamma-Tocotrienol as a hypocholesterolemic and antioxidant agent in rats fed atherogenic diets.* Lipids. 1993; 28: 1113–8.

Weinbrenner T, Schroder H, Escurriol V et al. *Circulating oxidized LDL is associated with increased waist circumference independent of body mass index in men and women.* Am J Clin Nutr. 2006; 83: 30–5; quiz 181–2.

Yasunari K, Maeda K, Nakamura M et al. *Oxidative stress in leukocytes is a possible link between blood pressure, blood glucose, and C-reacting protein.* Hypertension. 2002; 39: 777–80.

Young IS, McFarlane C, McEneny J. *Oxidative modification of triacylglycerol-rich lipoproteins.* Biochem Soc Trans. 2003; 31:1062–5.

Zalba G, San Jose G, Moreno MU et al. *Oxidative stress in arterial hypertension: role of NAD(P)H oxidase.* Hypertension. 2001; 38: 1395–9.

Zheng L, Nukuna B, Brennan ML et al. *Apolipoprotein A-I is a selective target for myeloperoxidase-catalyzed oxidation and functional impairment in subjects with cardiovascular disease.* J Clin Invest. 2004; 114: 529–41.

Chapter 4
Chronic Inflammation in the Metabolic Syndrome: Emphasis on Adipose Tissue

Rosário Monteiro

Contents

Abstract The increasing incidence of the metabolic syndrome and related pathologies is disturbing. The activation of inflammatory pathways, used normally as host defence response, reminds us of the life threatening nature of this condition. Although initially directed at relieving the causes of the syndrome, as it progresses the inflammatory response becomes maladaptive and seems to largely contribute to the pathological outcomes associated with the metabolic syndrome. A single cause for the activation of inflammation is difficult to establish, as is a timeline of events related to the deterioration of metabolic homeostasis. Apparently, metabolic overload evokes several stress reactions, such as oxidative, inflammatory, organelle and

R. Monteiro (✉)
Department of Biochemistry (U38/FCT), Faculty of Medicine, University of Porto,
Al. Prof. Hernâni Monteiro, 4200-319 Porto, Portugal
e-mail: rosariom@med.up.pt

R. Soares, C. Costa (eds.), *Oxidative Stress, Inflammation and Angiogenesis in the Metabolic Syndrome,* DOI 10.1007/978-1-4020-9701-0_4,

cell hypertrophy stresses, generating vicious cycles that amplify each other leading to dysfunction. The difficulty in the management of the syndrome is linked to its multifactorial nature where environmental, genetic and psychosocial factors play a role.

Keywords Adipokines · Adipose tissue dysfunction · Endoplasmic reticulum stress · Hypertrophy · Macrophages

4.1 The Metabolic Syndrome

The incidence of the metabolic syndrome is increasing in developed and developing countries. This constellation of disturbances including glucose intolerance, central obesity, dyslipidaemia (hypertriglyceridemia, elevated non-esterified fatty acids (NEFA) and decreased high density lipoprotein (HDL) cholesterol) and hypertension can present in several forms, according to the combination of the different components of the syndrome (Eckel et al. 2005). It is widely accepted that the metabolic syndrome largely increases the risk for the development of cardiovascular disease, type 2 diabetes and cancer (Eckel et al. 2005). However, intense debate is ongoing regarding the causes for the onset of the metabolic disturbances that constitute the syndrome and there have been several attempts to define it that have drawn special attention to one or another component. For example, the American Association of Endocrinology does not consider obesity as a component and highlights the importance of insulin resistance to the syndrome (Einhorn et al. 2002). The initial definition of the World Health Organization also considered insulin resistance a key feature of the metabolic syndrome (Alberti and Zimmet 1998; Balkau and Charles 1999), while the more recent National Cholesterol Education Program (NECP):Adult Treatment Panel III (ATP III) definition adds equal weight to any of the components of the syndrome: glucose intolerance, obesity, hypertension and dyslipidaemia (NECP:ATPIII 2001).

Several explanations have been proposed to explain the origin of the metabolic syndrome. While some consider an initial insulin resistant state progressing to the other components, others think that obesity is the main initiator of the syndrome (Laclaustra et al. 2007; Mittra et al. 2008). More recently, the chronic low-grade inflammatory condition that accompanies the metabolic syndrome has been implicated as a major player both in the installation of the metabolic syndrome and its associated pathophysiological consequences (Wellen and Hotamisligil 2005; Hotamisligil 2006; de Ferranti and Mozaffarian 2008).

4.2 The Inflammatory Response

Inflammation is a physiological response of the organism to harmful stimuli, namely infection or tissue injury. The response provided is usually set towards the re-establishment of homeostasis. It involves the coordinated action of many cell types and mediators, whose intervention depends on the nature of the initial stimulus (Lawrence and Gilroy 2007; Medzhitov 2008). The normal acute inflammatory response involves the delivery of plasma components and leucocytes to the site of insult and is initiated by tissue-resident macrophages and mast cells, in the case of innate response to infection, leading to a production of different types of inflammatory mediators (chemokines, cytokines, vasoactive amines, eicosanoids and products of proteolytic cascades) (Lawrence and Gilroy 2007; Medzhitov 2008). The stimulated endothelium allows extravasation of neutrophils and soluble components to the tissue, where they become activated releasing toxic agents and proteolytic enzymes to the extracellular milieu (Serhan 2007). If successful, the injurious agent is eliminated and inflammation resolution and tissue repair follow. This is achieved by switching the lipid mediators from pro-inflammatory (e.g. prostaglandins) to anti-inflammatory and pro-resolution ones (lipoxins, resolvins and protectins) and by the action of tissue-resident and newly recruited macrophages (Serhan 2007). Tissue leucocytes undergo apoptosis and are phagocytised by the macrophages that leave the inflamed site by lymphatic drainage (Bellingan et al. 2002). The apoptosis of inflammatory cells is a non-phlogistic physiological process of removing dead cells and is essential for the resolution of inflammation (Lawrence and Gilroy 2007; Serhan 2007) with the added advantage that after engulfing these apoptotic cells, macrophages acquire a phenotype conductive to resolution, releasing anti-inflammatory signals such as interleukin (IL)-10 and transforming growth factor β (TGFβ) (Huynh et al. 2002).

However, if the neutralization and removal of the noxious stimuli, or even if the clearance of apoptotic inflammatory cells from the inflamed tissue fail, the inflammatory process persists and a state of chronic inflammation or auto-immunity may arise with different agents being recruited, namely T lymphocytes and the development of lymphoid infiltrates in the tissue (Lawrence and Gilroy 2007).

4.2.1 Inflammation and the Metabolic Syndrome

It has been discussed that the inflammatory state that accompanies the metabolic syndrome does not completely fit into the classical definition of acute or chronic inflammation, in that it is not accompanied by infection and no massive tissue injury seems to have taken place. Furthermore, the dimension of the inflammatory activation is not as large and so it is often called 'low grade' chronic inflammation. Other researchers have attempted to name this inflammatory state as 'metaflammation', meaning metabolically-triggered inflammation (Hotamisligil 2006) or even 'para-inflammation' as a term to define an intermediate state between basal and inflammatory states (Medzhitov 2008). Whatever the term used, the inflammatory process that characterizes the metabolic syndrome has its own unique features. On the other hand, its causes are far from being fully understood.

Tissue malfunction or homeostatic imbalance of one or several physiological systems seems to be in the basis of the inflammatory process, but it is sometimes difficult to understand how this would elicit a host defence mechanism. Instead of being beneficial, it seems to progress to a deregulated state and is implicated in the worsening of the condition. It has even been questioned whether there is any physiological counterpart for this inflammatory response (Medzhitov 2008). It is possible that the initial response provides short-term benefits but that in a chronic phase it becomes maladaptive. Indeed, inducible adaptive changes generally occur at the expenses of many other physiological processes and, therefore, cannot be sustained without adverse side effects caused by the decline in the affected functions (Lawrence and Gilroy 2007; Medzhitov 2008).

4.2.2 Signalling Pathways in Inflammation

Most known signalling pathways related to inflammation involve members of the IL-1 and tumor necrosis factor (TNF) receptor families and the Toll-like microbial pattern recognition receptors (TLR) (Lawrence and Gilroy 2007). The first two receptors are activated by IL-1 and TNFα, respectively, the typical cytokines released upon tissue injury. TLR recognize microbial patterns and are thus termed pattern recognition receptors. These receptors represent a germline encoded nonself recognition system that, when activated, triggers the inflammatory response (Lawrence and Gilroy 2007). Activation of either one of these receptors recruits adaptor proteins linked to the regulation of cell survival, that further recruit signalling proteins that lead to the activation of mitogen-activated protein kinases (MAPK) (e.g. jun N-terminal kinase (JNK) and p38 MAPK) and inhibitor of nuclear factor κB kinases (IKK) (Chang and Karin 2001; Kyriakis and Avruch 2001). The MAPK lead to the phosphorylation of transcription factors, such as activator protein 1 (AP-1) (Karin 1995) and cAMP response element-binding (CREB) proteins (Park et al. 2005), which bind to promoters of pro-inflammatory genes activating their transcription. These kinases are also able to induce post-transcriptional mechanisms to regulate pro-inflammatory gene expression (Chen et al. 1998; Dean et al. 2004; Winzen et al. 2004). IKK are responsible for the activation of nuclear factor κB (NFκB) transcription factor (Ghosh and Karin 2002), recognized as a central player in inflammatory and immune responses (Li and Verma 2002; Bonizzi and Karin 2004). Both MAPK and IKK induce expression of IL-1 and TNFα, meaning that they are able to amplify the immune response. Among the targets of MAPK and IKK are also IL-6 and IL-12 (O'Shea et al. 2002).

Activation of cytokine receptors or TLR results in activation of phosphoinositide-3-kinases, which may themselves activate other kinases, like Akt (Martin et al. 2003). The coordinated action of these signalling cascades results in the initiation and maintenance of the inflammatory response.

4.2.3 Toll-Like Receptors

The best known receptors involved in the innate immune response are the TLR. Indeed, these receptors are widely expressed in cells of the immune system, such as macrophages, as well as in organ parenchyma in epithelial and endothelial cells

(Wolowczuk et al. 2008). It is increasingly recognized that they do also participate in adaptive immune response, given their expression in B lymphocytes, mast cells, T lymphocytes and dendritic cells (Wolowczuk et al. 2008).

The two most widely studied TLR are TLR2 and TLR4, which are activated by bacterial lipoproteins and by lipopolysaccharide, respectively (Wolowczuk et al. 2008). The engagement of either receptor leads to translocation of NFκB to the nucleus. In addition to their early recognized role in immunity, a participation in the regulation of metabolism is also being attributed to TLR (Wolowczuk et al. 2008). It has been shown that these receptors may be activated by specific types of lipids. It had already been demonstrated that the fatty acid moiety of TLR ligands was essential for their stimulation (Raetz 1990) and this led to the investigation of its possible activation by different sorts of lipids. Thus, it was discovered that saturated fatty acids activate both TLR2 and TLR4 and, instead, unsaturated fatty acids inhibit TLR-mediated signalling and gene expression (Lee et al. 2001). This was also demonstrated with diet-derived saturated fatty acids, which inhibited TLR-mediated expression of IL-6 and TNFα whereas unsaturated fatty acids had no effect alone, but inhibited the saturated fatty acid-induced increase in TNFα expression (Shi et al. 2006).

Activation of TLR2 has been studied with regard to the development of atherosclerotic plaque, since it seems that in lipid rafts TLR2 binds to CD36 facilitating the transfer of fatty acids into the cells and contributing to atherosclerosis progression (Hoebe et al. 2005).

Although it has become evident that adipocytes are involved in innate immunity, only recently was the presence of TLR in these cells described. TLR4 has been found in murine preadipocytes 3T3-L1 and both TLR2 and TLR4 have been shown to be present and functional in human subcutaneous adipocytes (Lin et al. 2000; Bes-Houtmann et al. 2007). Activation of TLR results in synthesis of pro-inflammatory factors such as TNFα, IL-6 and chemokines (Lin et al. 2000; Bes-Houtmann et al. 2007; Poulain-Godefroy and Froguel 2007). Considering the implication of the increase in circulatory NEFA in adipose tissue dysfunction, it is very likely that the activation of TLR takes place in hyperlipidemic states, resulting in amplified inflammation and contributing to the development or aggravation of the metabolic syndrome (Wolowczuk et al. 2008).

4.2.4 Advanced Glycation End-Products

The involvement of advanced glycation end-products (AGE) in the pathogenesis of the metabolic syndrome is being given more and more attention. These molecules, formed in the context of oxidative stress and hyperglycaemia, constitute a heterogeneous group of substances with different physicochemical properties (Xanthis et al. 2007). The most well studied AGE in vivo are carboxymethyl lysine and pentosidine, which have been found increased in plasma and tissues in hyperglyceamia and been attributed a role in diabetic vascular atheromatosis (Chuyen 2006). They are able to crosslink with endothelium, basement membrane, matrix proteins, cellular membranes, DNA, intracellular proteins, and lipoproteins (Schmidt et al. 1995; Vlassara et al. 1995). A role of these substances in inflammation linked to the metabolic syndrome has been advanced. They act on cell receptors of AGE (RAGE), activating NFκB via MAPK and leading to the transcription of pro-inflammatory factors (Lin 2006). The result is increased production of endothelin and plasminogen activation inhibitor-I, enhanced accumulation and adherence of monocytes and platelets to the endothelium through increased vascular endothelial cell adhesion molecule expression and inflammation. Experimental evidence shows that AGE may also elicit innate immune responses (Liliensiek et al. 2004).

4.3 The Adipose Tissue

The primarily known functions of adipose tissue include heat insulation, mechanical cushioning and storage of triacylglycerols (Laclaustra et al. 2007).

Furthermore, adipose tissue secretes endocrine, paracrine and autocrine active substances in response to different stimuli (Kershaw and Flier 2004; Lafontan 2005). Some are specific to adipocytes, such as the adipokines leptin and adiponectin, and other may be produced by several cell types in the adipose tissue and include inflammatory cytokines (TNFα, IL-6), chemokines (monocyte chemoattractant protein-1), acute phase reactants, components of the alternative complement system, eicosanoids, as well as molecules with anti-inflammatory properties (Bays et al. 2008).

The metabolic deregulations that arise with obesity are mainly the reflection of excess energy intake, in face of expenditure, the demand being above the capacity of the organism to store the excess of energy (Laclaustra et al. 2007). The homeostatic role of the adipose tissue assures that after a meal fatty acids, mainly derived from triglyceride-rich lipoproteins, are avidly taken up by adipocytes, and during fasting or increased expenditure periods, the fatty acids are easily released to the blood stream. This is usually achieved by the coordinated action of hormones, catecholamines and insulin being the chief regulators of this balance (Laclaustra et al. 2007). Not only is the adipocyte metabolism altered by these hormones, but these hormones also regulate blood flow into the adipose tissue through the regulation of vascular tone, such that there is increased blood supply when the organ is metabolically active (Frayn et al. 2003).

4.3.1 Adipose Tissue in the Metabolic Syndrome

The role of the adipose tissue in the pathophysiology of the metabolic syndrome has received much attention in the last few years. Epidemiological evidence linking obesity to the predisposition to develop cardiovascular disease, type 2 diabetes and cancer has encouraged the study of adipose-based mechanisms in the metabolic syndrome (Laclaustra et al. 2007). Furthermore, the recognition of the adipose tissue as a true endocrine organ (Kershaw and Flier 2004; Lafontan 2005), rather than a passive storage for energy, with the description of its ability to produce inflammation- active molecules (Hotamisligil et al. 1993) has allowed beginning to understand the links between the adipose tissue and the other pathological manifestations of the metabolic syndrome. However, it was not until very recently that a paradigm shift has occurred in the way that we understand the association of the adipose tissue and metabolic disease. If the classical views have considered excess adipose tissue essentially detrimental for all metabolic outcomes, there is now the clear recognition that adipose tissue pathogenicity markedly differs according to location of adipose tissue deposition (visceral or subcutaneous) (Lafontan and Berlan 2003). Currently, there is even the suggestion that subcutaneous fat depots are protective and that the lack of adipose tissue, as evident in lipodystrophy (Hegele et al. 2007), may equally lead to the development of metabolic syndrome. For this reason, some authors now consider adipocyte dysfunction as the instigator of the main metabolic disturbances that constitute the metabolic syndrome and lead to the risk of disease (Laclaustra et al. 2007; Sethi and Vidal-Puig 2007; de Ferranti and Mozaffarian 2008).

4.3.2 Common Characteristics of Metabolic and Immune Systems

Metabolic and immune pathways have close evolutionary paths and are known to influence each other. In the presence of inflammatory stimuli, the anabolic messages are blunted in favour of a catabolic state usually needed to provide energy for the intervention of the immune system (Hotamisligil 2006). In the same manner, the nutritional and metabolic status largely influences the capability of an individual to built up an appropriate immune response, an ability that becomes threatened both in under or overnutriton (Hotamisligil 2006).

The liver and the adipose tissue possess chief roles in the management of energy requirements and in the storage of energy excess. Both organs are in close proximity to immune cells and there is evidence of mutual influence (Hotamisligil 2006). Regarding the adipose tissue, it has been suggested that adipose and immune organs may share a common ancestral progenitor (Caspar-Bauguil et al. 2005).

The biology of macrophages and adipocytes is many times overlapping. Both cell types express common genes such as the adipocyte-characteristic fatty acid binding protein 4 and peroxisome proliferator activated receptorγ (PPARγ) as well as macrophage-characteristic proteins such as inflammatory cytokines and matrix metalloproteinases (Hotamisligil et al. 1993; Tontonoz et al. 1998; Bouloumie et al. 2001; Makowski et al. 2001). Their functional capability is sometimes also common. For example, the ability of macrophages to take up lipids is evident in the formation of foam cells during the atherosclerotic process (Takahashi et al. 2002). Additionally, it has been shown that preadipocytes may phagocytise foreign particles, or have microbicidal properties, apart from being able to differentiate into macrophages when cultured in appropriate media (Cousin et al. 1999; Charriere et al. 2003). Also of importance is the participation of both cell types in the immune response: the macrophages directly killing pathogens and secreting cytokines and chemokines, whereas the adipocytes supply the inflammatory cells for energy and lipid mediators to support inflammatory response (Mattacks et al. 2004). The co-localization of adipocytes and macrophages in obesity has been set into attention with the work of Weisberg (Weisberg et al. 2003). It has also been shown that most of the inflammatory cytokine production from the adipose tissue of obese individuals is derived from inflammatory rather than adipose cells (Hotamisligil 2006). This suggests that these cells have paramount importance in the genesis of obesity complications.

4.3.3 The Presence of Inflammatory Cells in the Adipose Tissue

Several differences have been encountered between sites of adipose tissue deposition both in the type and number of inflammatory cells that they may host and in the production and/or integration of metabolic and inflammatory signals.

The presence of macrophages in the adipose tissue was reported a few years ago (Weisberg et al. 2003). Since then, it has been specified that these cells surround dead or dying adipocytes forming aggregates that have been termed crown-like structures (Cinti et al. 2005) and the proposed reason for their presence is prevention of the potential cytotoxicity arising from the spillage of the large fat droplet of the dying cell. For each dead adipocyte, several macrophages are recruited (Strissel et al. 2007), implying a large amplification of the inflammatory response since they are thought to locally produce inflammatory mediators. However, it appears that this promotion of inflammation by macrophages is only transient. Macrophages are usually involved in the recovery of inflammatory response, contributing to resolution, and in tissue remodelling (Bourlier et al. 2008). In this stage, probably after having phagocytised cell debris of apoptotic cells, it seems that macrophages engage an anti-inflammatory program secreting IL-10, IL-4 and TGFβ (Lawrence and Gilroy 2007; Strissel et al. 2007; Bourlier et al. 2008). Nevertheless, phagocytosis of necrotic cells leads to the expression of pro-inflammatory cytokines by the macrophages (Barker et al. 1999).

Apart from the presence of macrophages, lymphocytes have also been found in the adipose tissue, with differential distribution according to location. The lymphocytes found in epididimal adipose tissue, as in the liver, mainly belong to the innate immune system (Caspar-Bauguil et al. 2005). This suggests that, as for the liver, this is a site for early encounter of antigens, which makes sense for visceral fat, given the close proximity with antigens arriving from the gastrointestinal system in mesenteric adipose tissue. Furthermore, it has also been proposed that these cells may work in the primary defence through the recognition of self-antigens exposed by stressed, damaged and/or transformed cells of the host tissue (Jameson et al. 2004). On the other hand, in inguinal (subcutaneous) adipose tissue lymphocyte features resemble those of lymph nodes, thus belonging to the adaptive immune system (Caspar-Bauguil et al. 2005).

4.3.4 Adipocyte Hypertrophy, Dysfunction and Inflammation

The reasons for the appearance of macrophages in the adipose tissue are not fully known. It has been proposed that the activation of inflammatory signalling pathways or the production of inflammatory chemokines by adipose tissue cells is a stimulus for their recruitment (Neels and Olefsky 2006; Weisberg et al. 2006). In this regard, it has been shown that circulatory inflammatory cytokines (TNFα, IL-6, and C-reactive protein) are positively correlated with adipocyte size and, conversely, the plasma concentration of adiponectin is decreased with increasing cell size (Bahceci et al. 2007), suggesting that adipocyte hypertrophy may lie at the bottom of this question.

In hypertrophic cells, both mechanic and hypoxic stresses are increased (Gregor and Hotamisligil 2007). Either one of these disturbances may have impact on organelle dysfunction (see below) with consequent triggering of pro-inflammatory cytokine production and activation of pro-apoptotic pathways. The proper disposure of these apoptotic cells requires the intervention of macrophages, a massive recruitment of these cells being required since in histological sections of adipose tissue from obese animals it is possible to observe a large number of macrophages associated to only one dying adipocyte (Cinti et al. 2005; Strissel et al. 2007).

It is important to notice that, in resemblance to what happens in other cell types, it is likely that if apoptotic cells are not properly removed by macrophages, spillage of apoptotic vesicle content may exacerbate the inflammatory reaction (Lawrence and Gilroy 2007).

Knowing the correlation of adipocyte hypertrophy with inflammation and dysfunction, we have demonstrated that bigger adipocytes are more liable to rupture when exposed to common physical forces (Monteiro et al. 2006). Within the enclosed visceral cavity, sudden pressure variations occur with physical movements that may jeopardize the physical integrity of the already stressed cell membrane of hypertrophic adipocytes (Cobb et al. 2005). In obese individuals, the intra-abdominal pressure is even higher due to expanding adipose tissue (Lambert et al. 2005). It is, thus, very likely that necrotic rupture of adipocytes within the visceral cavity may contribute to amplify the call for macrophages to the adipose tissue, representing an important link between visceral obesity and inflammation.

Moreover, it has been proposed that visceral adipose tissue growth is mainly due to hypertrophy, while in other locations there may be mainly growth through hyperplasia (de Ferranti and Mozaffarian 2008). The physical constrains presented inside the abdominal cavity may halt adipogenic differentiation of adipose tissue precursors reducing the number of competent cells to accumulate excessive energy ingested. As a matter of fact, it has been shown that stretching inhibits adipocyte differentiation (Tanabe et al. 2004). On the other hand, having a smaller number of adipocytes results in at least two possible outcomes. Initially, the hypertrophic growth of the tissue may occur but later the buffering capacity of energy excess becomes exceeded, culminating in ectopic fat accumulation and dyslipidaemia. Indeed, it has been demonstrated that large adipocytes are a feature of insulin resistance (Jernas et al. 2006) and that the insulin-sensitizing anti-inflammatory thiazolidinediones have a pro-adipogenic effect, resulting in a higher capacity of fat storage by the adipose tissue (Sethi and Vidal-Puig 2007).

4.3.5 Adipose Tissue Dysfunction Leading to Dyslipidaemia

The most prominent sign of adipose tissue failure is the increase in circulating concentration of NEFA (Laclaustra et al. 2007). This reflects the inability of the tissue to buffer the excess nutrient intake and is related to the dyslipidemic state that is typical of the metabolic syndrome. When overload becomes present, the liver increases the production of apo-B containing particles that carry triacylglycerols to the adipose tissue resulting in low density lipoprotein (LDL) formation (Parhofer and Barrett 2006). This occurs in visceral adipose tissue very efficiently, but this depot is also more capable of releasing lipids in times of requirement. The subcutaneous adipose tissue has usually a much larger capacity to store lipids given its usually larger size. It seems that when the storage capacity of the subcutaneous adipose tissue is exhausted, energy starts to be accumulated in the visceral compartment (Romanski et al. 2000). This may explain why subcutaneous adipose tissue is considered protective in terms of metabolic syndrome and why men, who for genetic and hormonal reasons possess a smaller subcutaneous adipose tissue compartment, achieve earlier the limit of that depot and begin using the restricted capacity of the visceral depot. When the capacity of both locations is overwhelmed, the conversion of very low density lipoprotein (VLDL) or similar particles is delayed and hypertriglyceridemia originates (Laclaustra et al. 2007). Furthermore, neighbouring tissues start being used for lipid accumulation (e.g. liver, muscle, pancreas and heart) (Sethi and Vidal-Puig 2007). As these organs are not able to store lipids without harm to their functions, lipotoxicity may be the result culminating, in the case of muscle, liver and pancreas, in insulin resistance.

4.4 Insulin Signalling

When activated by insulin, the insulin receptor, which belongs to the family of tyrosine-kinase receptors, catalyses the phosphorylation of docking proteins to mediate signalling. The first downstream event in the cascade is tyrosine phosphorylation of insulin receptor substrate proteins (IRS-1 and IRS-2) (Pirola et al. 2004). This event is critical for proper insulin signalling and is often attenuated in systemic insulin resistance. Under normal circumstances IRS phosphorylation leads to the activation of two main signalling pathways: the phosphatidylinositol-3-kinase(PI3K)-protein kinase B(PKB)/Akt pathway, responsible form most of the metabolic actions of insulin (e.g. translocation of the glucose transporter to the membrane for glucose internalization, glycogen synthesis), and Ras-MAPK pathways that cooperate with PI3K to control cell growth and differentiation (Pirola et al. 2004).

In the insulin resistant state, IRS-1 phosphorylation in tyrosine does not take place, cell glucose uptake stops, glucose is retained in the extracellular space and hyperglycaemia occurs, which in turn stimulates insulin secretion by pancreatic β cells (Chitturi et al. 2002). Once the pancreas is depleted and can no longer compensate for hyperglycaemia, type 2 diabetes develops.

Besides the regulation of carbohydrate metabolism, the insulin action is also very important to control lipid storage and mobilization. It inhibits triglyceride hydrolysis and free fatty acid release from adipose tissue (Kitamura et al. 1999) through PI3K after IRS-1 phosphorylation with activation of phosphodiesterase and, as a consequence, cAMP degradation and depletion. With the decrease of cAMP, protein kinase A (PKA) is not stimulated, attenuating hormone sensitive lipase (HSL) activation (Anthonsen et al. 1998). As a consequence of insulin resistance, cAMP remains high in the adipose tissue, activating PKA which phosphorylates HSL and perilipin, allowing triacylglycerol degradation and NEFA release into the blood. The lipogenic and anti-lipolytic effects of insulin are coordinated by the PI3K-mediated effects of insulin on sterol regulatory element binding protein (SREBP), a transcription factor that plays an essential role in the activation of various genes involved in lipogenesis (e.g. acetyl-CoA carboxylase, fatty acid synthase, glycerol-3 phosphate acetyltransferase) and in VLDL excretion (Taskinen 2003). Hence, in the absence of insulin activity, all these genes are repressed, and so is lipogenesis.

The actions of insulin in the liver are mediated by IRS-2 tyrosine phosphorylation (Previs et al. 2000), which through PI3K/PKB/Akt phosphorylates and inactivates glycogen synthase kinase-3, which stops inhibiting glycogen synthase (Cross et al. 1995) resulting in increased glycogen synthesis in the liver. Liver insulin resistance results in opposite effects, not only decreasing glycogen synthesis but also increasing glycolysis, glyconeogenesis and glucose release into the circulation. In addition, insulin stimulates the expression of lipogenic genes determining the synthesis of fatty acids in the liver, an action that depends on IRS-2 and SREBP activation (Lopez et al. 1996).

4.4.1 Adipose Tissue-Derived Molecular Mechanisms of Insulin Resistance

Both the hypertriglyceridaemia that occurs during obesity or adipose tissue dysfunction and overproduction of cytokines by the inflamed adipose tissue contribute to the impairment of insulin signalling. It is known that several kinases phosphorylate the IRS in serine residues (Rudich et al. 2007). This phosphorylation in serine prevents IRS activation by the insulin receptor, blunts downstream signalling and facilitates the degradation of IRS protein.

There is compelling evidence showing that exposure of adipocytes to several types of stressors (oxidative stress, inflammatory cytokines, and elevated concentrations of fatty acids) induces cellular responses mediated by cellular kinases, including MAPK (p38MAPK, JNK and extracellular signal-regulated kinase), IKKβ, mammalian target of rapamycin (mTOR) and various conventional and atypical protein kinases C (PKC).

Some of these kinases involved in stress-sensing work in concert to stop the toxic increase of body energy stores, being related to the impairment of insulin action through the stimulation of IRS serine phosphorylation, but also often activate targets related to the inflammatory response (Hotamisligil 2006). The three main kinases that have been related to this inactivation of IRS are JNK, IKK and PKC (Griffin et al. 1999; Aguirre et al. 2000; Gao et al. 2002). As mentioned above, they exert powerful effects on pro-inflammatory gene expression, through activation of AP-1 complexes and NFκB (Baud and Karin 2001).

In obesity, there is a remarkable increase in JNK activity in the adipose tissue and liver (Hirosumi et al. 2002), which is thought to be due to exposure to NEFA and cytokines, such as TNFα, or due to cellular response to organelle stress (Aguirre et al. 2000; Ozcan et al. 2004; Wellen and Hotamisligil 2005). Its involvement in insulin resistance has been well demonstrated in genetic mice models of JNK1 deficiency that are protected from obesity induced JNK activation, IRS-1 serine phosphorylation, insulin resistance, fatty liver and diabetes (Hirosumi et al. 2002; Tuncman et al. 2006). Furthermore, blockade of JNK activity in models of obesity and diabetes improves systemic glucose homeostasis and insulin sensitivity, as well as atherosclerosis (Kaneto et al. 2004; Ricci et al. 2004; Liu and Rondinone 2005).

Metabolic dysfunction due to metabolic overload also seems to be mediated through IKKβ. The reduction of IKKβ expression partly protects mice from obesity-induced insulin resistance, and the inhibition of this kinase achieved by high doses of salycilates has been shown to improve insulin sensitivity in humans and other experimental models (Yuan et al. 2001; Hundal et al. 2002).

PKC has also been shown to constitute an important interface between metabolic deregulation, inflammation and insulin resistance. This kinase (isoform θ in skeletal muscle and δ in the liver) can be activated by fatty acid metabolites that accumulate due to metabolic pathway burden such as fatty acyl CoA and diacylglycerol, leading to inhibitory serine phosphorylation of IRS and attenuation of insulin signalling (Yu et al. 2002; Boden et al. 2005). Furthermore, PKCθ is known to activate IKK and might contribute to insulin resistance and amplification of inflammation.

4.5 Organelle Stresses in the Adipose Tissue as a Link to Inflammation

Adipose tissue in obesity is burdened with several types of insults such as inflammation, hypoxia, oxidative stress and mechanical stress attributable to hypertrophy. These insults cumulatively result in organelle dysfunction, particularly in mitochondria and the endoplasmic reticulum (ER).

The ER is a cytosolic organelle that participates in the regulation of lipid, glucose, cholesterol, and protein metabolism, apart from being the site of triglyceride droplet formation. Cells with high secretory activity, like the hepatocytes and pancreatic β cells, adapt and expand their ER to meet the demand (Federovitch et al. 2005).

As recognized in the last few years, the adipocyte is an endocrine cell that undergoes striking transformation during its life cycle. Furthermore, in the obese, the adipocyte may be especially challenged, given that it is required to secrete large amounts of substances and synthesise lipids. Under such conditions, ER function may be impaired leading to the accumulation of misfolded or unfolded proteins in its lumen. Under metabolic challenge, the same may occur in hepatocytes and pancreatic β cells. Indeed, ER stress was recently shown to be present in obese, insulin-resistant tissues in experimental models. Notably, this stress was most prominent in the adipose tissue and contributed to its dysfunction (Ozcan et al. 2004; Hotamisligil 2006). As a way to cope with it, the stressed endoplasmic reticulum engages the unfolded protein response (UPR). Other stresses, such as glucose and energy deprivation, inhibition of protein glycosylation or imbalance of ER calcium levels may elicit the UPR. The UPR functions via signalling through three branches, denoted for the three stress-sensing proteins found in the ER membrane: PKR-like eukaryotic initiation factor 2a kinase (PERK), inositol-requiring enzyme-1 (IRE-1), and activating transcription factor-6 (ATF-6) (Gregor and Hotamisligil 2007). Their activation depends upon the stress induced and graded responses are processed, initially through attenuation of the cellular workload (decreasing protein translation, clearance and degradation of excess proteins from the ER lumen), repair (induction of an antioxidant response and of chaperone transcription to assist with the unfolded proteins) and ER biogenesis, towards recovery and survival of the cell. However, if the ER stress is not relieved, the UPR may also induce cell death via apoptosis. JNK, after activation by IRE-1, is an important effector in this action (Wu and Kaufman 2006; Zhao and Ackerman 2006) and, apart from this role it may lead to a variety of other downstream effects depending on the cellular context, such as cell survival, inflammation, and insulin resistance. The activation of inflammation by the UPR also depends upon the IKK-NFκB pathway, also through IRE-1α, resulting in increases TNF-α and IL-6 production, further supporting its contribution to insulin resistance (Deng et al. 2004; Hu et al. 2006; Shoelson et al. 2006). The NFκB pathway may also be activated through PERK signalling during the UPR (Jiang et al. 2003; Deng et al. 2004; Wu et al. 2004). Finally, as mentioned, adipocyte death may be a contributor to the inflammation in obese adipose tissue (Cinti et al. 2005), and ER stress may play a role in this death via its ability to engage apoptotic pathways. It is also important to mention that, although having been demonstrated to be the most affected, adipose tissue is not the only affected by ER stress, as hepatocytes, cardiomyoblasts, pancreatic β-cells, and macrophages (Kharroubi et al. 2004; Borradaile et al. 2006a; Borradaile et al. 2006b; Karaskov et al. 2006; Wei et al. 2006; Gregor and Hotamisligil 2007) may also present with this disturbance.

Obesity is associated with deregulated lipid and carbohydrate metabolism. An increase in either one of these substrates will also increase the demand on the mitochondria and the utilization of the electron transport chain (Rudich et al. 2007). As in metabolically active tissues undergoing increased demand, there is usually relative hypoxia, together with the increased need for nutrient oxidation. This generates unusual amounts of reactive oxygen species. Oxidative stress activates kinases like

JNK, p38 MAPK and IKK that may directly interfere with insulin signalling or indirectly via induction of NFκB and increased cytokine production (Qatanani and Lazar 2007).

Furthermore, the enhanced flux through alternative metabolic pathways leads to the generation of other metabolites that may themselves be related to impairment of insulin function and inflammation. Examples include fatty acyl CoA, diacylglycerol, which activate PKC serine kinases, and ceramide, which can undergo phosphorylation to ceramide-1-phosphate and promote inflammation or originate secondary metabolites with the same effect. Ceramide may also activate protein phosphatase 2A, which inactivates PKB/Akt and attenuates insulin response, contributing to insulin resistance (Wymann and Schneiter 2008).

4.6 Conclusion

It has become evident that the inflammatory condition that is associated with obesity and overweight plays an important part in the aetiology of the metabolic syndrome and largely contributes to the related pathological outcomes. From the reasons here presented, it seems likely that adipocyte dysfunction lies at the bottom of this question, its homeostatic functions being overwhelmed due to metabolic overload. From then on, several vicious cycles exacerbate the disturbances and lead to an inflammatory response that initially aims at restoring homeostasis and later contributes to disease. This maladaptive inflammatory reaction results in alterations in physiological processes and is sustained at the expense of the decline in tissue function reflecting in adverse side effects. The awareness of the importance of inflammation, mainly that originated from dysfunctional adipose tissue, in the metabolic syndrome, may help design new strategies to prevent of treat metabolic syndrome related diseases.

References

Aguirre V, Uchida T, Yenush L, Davis R and White MF. *The c-Jun NH(2)-terminal kinase promotes insulin resistance during association with insulin receptor substrate-1 and phosphorylation of Ser(307).* J Biol Chem. 2000; 275: 9047–54.

Alberti KG and Zimmet PZ. *Definition, diagnosis and classification of diabetes mellitus and its complications. Part 1: diagnosis and classification of diabetes mellitus provisional report of a WHO consultation.* Diabet Med. 1998; 15: 539–53.

Anthonsen MW, Ronnstrand L, Wernstedt C, Degerman E and Holm C. *Identification of novel phosphorylation sites in hormone-sensitive lipase that are phosphorylated in response to isoproterenol and govern activation properties in vitro.* J Biol Chem. 1998; 273: 215–21.

Bahceci M, Gokalp D, Bahceci S, Tuzcu A, Atmaca S and Arikan S. *The correlation between adiposity and adiponectin, tumor necrosis factor alpha, interleukin-6 and high sensitivity C-reactive protein levels. Is adipocyte size associated with inflammation in adults?* J Endocrinol Invest. 2007; 30: 210–4.

Balkau B and Charles MA. *Comment on the provisional report on the provisional report from WHO consultation. European group for the study of insulin resistance (EGIR).* Diabet Med. 1999; 16: 442–3.

Barker RN, Erwig L, Pearce WP, Devine A and Rees AJ. *Differential effects of necrotic or apoptotic cell uptake on antigen presentation by macrophages.* Pathobiology. 1999; 67: 302–5.

Baud V and Karin M. *Signal transduction by tumor necrosis factor and its relatives.* Trends Cell Biol. 2001; 11: 372–7.

Bays HE, Gonzalez-Campoy JM, Bray GA, Kitabchi AE, Bergman DA, Schorr AB, Rodbard HW and Henry RR.*Pathogenic potential of adipose tissue and metabolic consequences of adipocyte hypertrophy and increased visceral adiposity.* Expert Rev Cardiovasc Ther. 2008; 6: 343–68.

Bellingan GJ, Xu P, Cooksley H, Cauldwell H, Shock A, Bottoms S, Haslett C, Mutsaers SE and Laurent GJ. *Adhesion molecule-dependent mechanisms regulate the rate of macrophage clearance during the resolution of peritoneal inflammation.* J Exp Med. 2002; 196: 1515–21.

Bes-Houtmann S, Roche R, Hoareau L, Gonthier MP, Festy F, Caillens H, Gasque P, Lefebvre d'Hellencourt C and Cesari M. *Presence of functional TLR2 and TLR4 on human adipocytes.* Histochem Cell Biol. 2007; 127: 131–7.

Boden G, She P, Mozzoli M, Cheung P, Gumireddy K, Reddy P, Xiang X, Luo Z and Ruderman N. *Free fatty acids produce insulin resistance and activate the proinflammatory nuclear factor-kappaB pathway in rat liver.* Diabetes. 2005; 54: 3458–65.

Bonizzi G and Karin M. *The two NF-kappaB activation pathways and their role in innate and adaptive immunity.* Trends Immunol. 2004; 25: 280–8.

Borradaile NM, Buhman KK, Listenberger LL, Magee CJ, Morimoto ET, Ory DS and Schaffer JE. *A critical role for eukaryotic elongation factor 1A-1 in lipotoxic cell death.* Mol Biol Cell. 2006a; 17: 770–8.

Borradaile NM, Han X, Harp JD, Gale SE, Ory DS and Schaffer JE. *Disruption of endoplasmic reticulum structure and integrity in lipotoxic cell death.* J Lipid Res. 2006b; 47: 2726–37.

Bouloumie A, Sengenes C, Portolan G, Galitzky J and Lafontan M. *Adipocyte produces matrix metalloproteinases 2 and 9: involvement in adipose differentiation.* Diabetes. 2001; 50: 2080–6.

Bourlier V, Zakaroff-Girard A, Miranville A, De Barros S, Maumus M, Sengenes C, Galitzky J, Lafontan M, Karpe F, Frayn KN and Bouloumie A. *Remodeling phenotype of human subcutaneous adipose tissue macrophages.* Circulation. 2008; 117: 806–15.

Caspar-Bauguil S, Cousin B, Galinier A, Segafredo C, Nibbelink M, Andre M, Casteilla L and Penicaud L. *Adipose tissues as an ancestral immune organ: site-specific change in obesity.* FEBS Lett. 2005; 579: 3487–92.

Chang L and Karin M. *Mammalian MAP kinase signalling cascades.* Nature. 2001; 410: 37–40.

Charriere G, Cousin B, Arnaud E, Andre M, Bacou F, Penicaud L and Casteilla L. *Preadipocyte conversion to macrophage. Evidence of plasticity.* J Biol Chem. 2003; 278: 9850–5.

Chen CY, Del Gatto-Konczak F, Wu Z and Karin M.*Stabilization of interleukin-2 mRNA by the c-Jun NH2-terminal kinase pathway.* Science. 1998; 280: 1945–9.

Chitturi S, Abeygunasekera S, Farrell GC, Holmes-Walker J, Hui JM, Fung C, Karim R, Lin R, Samarasinghe D, Liddle C, Weltman M and George J. *NASH and insulin resistance: Insulin hypersecretion and specific association with the insulin resistance syndrome.* Hepatology. 2002; 35: 373–9.

Chuyen NV. *Toxicity of the AGEs generated from the Maillard reaction: on the relationship of food-AGEs and biological-AGEs.* Mol Nutr Food Res. 2006; 50: 1140–9.

Cinti S, Mitchell G, Barbatelli G, Murano I, Ceresi E, Faloia E, Wang S, Fortier M, Greenberg AS and Obin MS. *Adipocyte death defines macrophage localization and function in adipose tissue of obese mice and humans.* J Lipid Res. 2005; 46: 2347–55.

Cobb WS, Burns JM, Kercher KW, Matthews BD, James Norton H and Todd Heniford B. *Normal intraabdominal pressure in healthy adults.* J Surg Res. 2005; 129: 231–5.

Cousin B, Munoz O, Andre M, Fontanilles AM, Dani C, Cousin JL, Laharrague P, Casteilla L and Penicaud L. *A role for preadipocytes as macrophage-like cells.* FASEB J. 1999; 13: 305–12.

Cross DA, Alessi DR, Cohen P, Andjelkovich M and Hemmings BA.*Inhibition of glycogen synthase kinase-3 by insulin mediated by protein kinase B.* Nature. 1995; 378: 785–9.

de Ferranti S and Mozaffarian D. *The perfect storm: obesity, adipocyte dysfunction, and metabolic consequences.* Clin Chem. 2008; 54: 945–55.

Dean JL, Sully G, Clark AR and Saklatvala J. *The involvement of AU-rich element-binding proteins in p38 mitogen-activated protein kinase pathway-mediated mRNA stabilisation.* Cell Signal. 2004; 16: 1113–21.

Deng J, Lu PD, Zhang Y, Scheuner D, Kaufman RJ, Sonenberg N, Harding HP and Ron D. *Translational repression mediates activation of nuclear factor kappa B by phosphorylated translation initiation factor 2.* Mol Cell Biol. 2004; 24: 10161–8.

Eckel RH, Grundy SM and Zimmet PZ. *The metabolic syndrome.* Lancet. 2005; 365: 1415–28.

Einhorn D, Reaven GM and Cobin RH. *American College of Endocrinology position statement on the insulin resistance syndrome.* Endocr Pract. 2002; 9: 236–52.

Federovitch CM, Ron D and Hampton RY. *The dynamic ER: experimental approaches and current questions.* Curr Opin Cell Biol. 2005; 17: 409–14.

Frayn KN, Karpe F, Fielding BA, Macdonald IA and Coppack SW. *Integrative physiology of human adipose tissue.* Int J Obes Relat Metab Disord. 2003; 27: 875–88.

Gao Z, Hwang D, Bataille F, Lefevre M, York D, Quon MJ and Ye J. *Serine phosphorylation of insulin receptor substrate 1 by inhibitor kappa B kinase complex.* J Biol Chem. 2002; 277: 48115–21.

Ghosh S and Karin M. *Missing pieces in the NF-kappaB puzzle.* Cell. 2002; 109 Suppl: S81–96.

Gregor MF and Hotamisligil GS. *Thematic review series: Adipocyte Biology. Adipocyte stress: the endoplasmic reticulum and metabolic disease.* J Lipid Res. 2007; 48: 1905–14.

Griffin ME, Marcucci MJ, Cline GW, Bell K, Barucci N, Lee D, Goodyear LJ, Kraegen EW, White MF and Shulman GI. *Free fatty acid-induced insulin resistance is associated with activation of protein kinase C theta and alterations in the insulin signaling cascade.* Diabetes. 1999; 48: 1270–4.

Hegele RA, Joy TR, Al-Attar SA and Rutt BK. Thematic review series: Adipocyte Biology. *Lipodystrophies: windows on adipose biology and metabolism.* J Lipid Res. 2007; 48: 1433–44.

Hirosumi J, Tuncman G, Chang L, Gorgun CZ, Uysal KT, Maeda K, Karin M and Hotamisligil GS. *A central role for JNK in obesity and insulin resistance.* Nature. 2002; 420: 333–6.

Hoebe K, Georgel P, Rutschmann S, Du X, Mudd S, Crozat K, Sovath S, Shamel L, Hartung T, Zahringer U and Beutler B. *CD36 is a sensor of diacylglycerides.* Nature. 2005; 433: 523–7.

Hotamisligil GS. *Inflammation and metabolic disorders.* Nature. 2006; 444: 860–7.

Hotamisligil GS, Shargill NS and Spiegelman BM. *Adipose expression of tumor necrosis factor-alpha: direct role in obesity-linked insulin resistance.* Science. 1993; 259: 87–91.

Hu P, Han Z, Couvillon AD, Kaufman RJ and Exton JH. *Autocrine tumor necrosis factor alpha links endoplasmic reticulum stress to the membrane death receptor pathway through IRE1alpha-mediated NF-kappaB activation and down-regulation of TRAF2 expression.* Mol Cell Biol. 2006; 26: 3071–84.

Hundal RS, Petersen KF, Mayerson AB, Randhawa PS, Inzucchi S, Shoelson SE and Shulman GI. *Mechanism by which high-dose aspirin improves glucose metabolism in type 2 diabetes.* J Clin Invest. 2002; 109: 1321–6.

Huynh ML, Fadok VA and Henson PM. *Phosphatidylserine-dependent ingestion of apoptotic cells promotes TGF-beta1 secretion and the resolution of inflammation.* J Clin Invest. 2002; 109: 41–50.

Jameson JM, Sharp LL, Witherden DA and Havran WL. *Regulation of skin cell homeostasis by gamma delta T cells.* Front Biosci. 2004; 9: 2640–51.

Jernas M, Palming J, Sjoholm K, Jennische E, Svensson PA, Gabrielsson BG, Levin M, Sjogren A, Rudemo M, Lystig TC, Carlsson B, Carlsson LM and Lonn M. *Separation of human adipocytes by size: hypertrophic fat cells display distinct gene expression.* FASEB J. 2006; 20: 1540–2.

Jiang HY, Wek SA, McGrath BC, Scheuner D, Kaufman RJ, Cavener DR and Wek RC. *Phosphorylation of the alpha subunit of eukaryotic initiation factor 2 is required for activation of NF-kappaB in response to diverse cellular stresses.* Mol Cell Biol. 2003; 23: 5651–63.

Kaneto H, Nakatani Y, Miyatsuka T, Kawamori D, Matsuoka TA, Matsuhisa M, Kajimoto Y, Ichijo H, Yamasaki Y and Hori M. *Possible novel therapy for diabetes with cell-permeable JNK-inhibitory peptide.* Nat Med. 2004; 10: 1128–32.

Karaskov E, Scott C, Zhang L, Teodoro T, Ravazzola M and Volchuk A. *Chronic palmitate but not oleate exposure induces endoplasmic reticulum stress, which may contribute to INS-1 pancreatic beta-cell apoptosis*. Endocrinology. 2006; 147: 3398–407.

Karin M. T*he regulation of AP-1 activity by mitogen-activated protein kinases.* J Biol Chem. 1995; 270: 16483–6.

Kershaw EE and Flier JS. *Adipose tissue as an endocrine organ.* J Clin Endocrinol Metab. 2004; 89: 2548–56.

Kharroubi I, Ladriere L, Cardozo AK, Dogusan Z, Cnop M and Eizirik DL. *Free fatty acids and cytokines induce pancreatic beta-cell apoptosis by different mechanisms: role of nuclear factor-kappaB and endoplasmic reticulum stress.* Endocrinology. 2004; 145: 5087–96.

Kitamura T, Kitamura Y, Kuroda S, Hino Y, Ando M, Kotani K, Konishi H, Matsuzaki H, Kikkawa U, Ogawa W and Kasuga M. *Insulin-induced phosphorylation and activation of cyclic nucleotide phosphodiesterase 3B by the serine-threonine kinase Akt.* Mol Cell Biol. 1999; 19: 6286–96.

Kyriakis JM and Avruch J. *Mammalian mitogen-activated protein kinase signal transduction pathways activated by stress and inflammation.* Physiol Rev. 2001; 81: 807–69.

Laclaustra M, Corella D and Ordovas JM. *Metabolic syndrome pathophysiology: the role of adipose tissue.* Nutr Metab Cardiovasc Dis. 2007; 17: 125–39.

Lafontan M. *Fat cells: afferent and efferent messages define new approaches to treat obesity.* Annu Rev Pharmacol Toxicol. 2005; 45: 119–46.

Lafontan M and Berlan M. *Do regional differences in adipocyte biology provide new pathophysiological insights?* Trends Pharmacol Sci. 2003; 24: 276–83.

Lambert DM, Marceau S and Forse RA. *Intra-abdominal pressure in the morbidly obese.* Obes Surg. 2005; 15: 1225–32.

Lawrence T and Gilroy DW. *Chronic inflammation: a failure of resolution?* Int J Exp Pathol. 2007; 88: 85–94.

Lee JY, Sohn KH, Rhee SH and Hwang D. *Saturated fatty acids, but not unsaturated fatty acids, induce the expression of cyclooxygenase-2 mediated through Toll-like receptor 4.* J Biol Chem. 2001; 276: 16683–9.

Li Q and Verma IM. *NF-kappaB regulation in the immune system.* Nat Rev Immunol. 2002; 2: 725–34.

Liliensiek B, Weigand MA, Bierhaus A, Nicklas W, Kasper M, Hofer S, Plachky J, Grone HJ, Kurschus FC, Schmidt AM, Yan SD, Martin E, Schleicher E, Stern DM, Hammerling GG, Nawroth PP and Arnold B.*Receptor for advanced glycation end products (RAGE) regulates sepsis but not the adaptive immune response.* J Clin Invest. 2004; 113: 1641–50.

Lin L. *RAGE on the Toll Road?* Cell Mol Immunol. 2006; 3: 351–8.

Lin Y, Lee H, Berg AH, Lisanti MP, Shapiro L and Scherer PE. *The lipopolysaccharide-activated toll-like receptor (TLR)-4 induces synthesis of the closely related receptor TLR-2 in adipocytes.* J Biol Chem. 2000; 275: 24255–63.

Liu G and Rondinone CM. *JNK: bridging the insulin signaling and inflammatory pathway.* Curr Opin Investig Drugs. 2005; 6: 979–87.

Lopez JM, Bennett MK, Sanchez HB, Rosenfeld JM and Osborne TF. *Sterol regulation of acetyl coenzyme A carboxylase: a mechanism for coordinate control of cellular lipid.* Proc Natl Acad Sci USA. 1996; 93: 1049–53.

Makowski L, Boord JB, Maeda K, Babaev VR, Uysal KT, Morgan MA, Parker RA, Suttles J, Fazio S, Hotamisligil GS and Linton MF. *Lack of macrophage fatty-acid-binding protein aP2 protects mice deficient in apolipoprotein E against atherosclerosis.* Nat Med. 2001; 7: 699–705.

Martin M, Schifferle RE, Cuesta N, Vogel SN, Katz J and Michalek SM. *Role of the phosphatidylinositol 3 kinase-Akt pathway in the regulation of IL-10 and IL-12 by Porphyromonas gingivalis lipopolysaccharide.* J Immunol. 2003; 171: 717–25.

Mattacks CA, Sadler D and Pond CM. *Site-specific differences in fatty acid composition of dendritic cells and associated adipose tissue in popliteal depot, mesentery, and omentum and their modulation by chronic inflammation and dietary lipids.* Lymphat Res Biol. 2004; 2: 107–29.

Medzhitov R. *Origin and physiological roles of inflammation.* Nature. 2008; 454: 428–35.
Mittra S, Bansal VS and Bhatnagar PK. *From a glucocentric to a lipocentric approach towards metabolic syndrome.* Drug Discov Today. 2008; 13: 211–8.
Monteiro R, de Castro PM, Calhau C and Azevedo I. *Adipocyte size and liability to cell death.* Obes Surg. 2006; 16: 804–6.
NECP:ATPIII. *Executive summary of the National Cholesterol Education Program (NECP) Expert panel on Detection, Evaluation and Treatment of High Blood Cholesterol in Adults (Adult Treatment Panel III).* JAMA. 2001; 285: 2486–97.
Neels JG and Olefsky JM. *Inflamed fat: what starts the fire?* J Clin Invest. 2006; 116: 33-5.
O'Shea JJ, Gadina M and Schreiber RD. *Cytokine signaling in 2002: new surprises in the Jak/Stat pathway.* Cell. 2002; 109 Suppl: S121–31.
Ozcan U, Cao Q, Yilmaz E, Lee AH, Iwakoshi NN, Ozdelen E, Tuncman G, Gorgun C, Glimcher LH and Hotamisligil GS. *Endoplasmic reticulum stress links obesity, insulin action, and type 2 diabetes.* Science. 2004; 306: 457–61.
Parhofer KG and Barrett PH. *Thematic review series: patient-oriented research. What we have learned about VLDL and LDL metabolism from human kinetics studies.* J Lipid Res. 2006; 47: 1620–30.
Park JM, Greten FR, Wong A, Westrick RJ, Arthur JS, Otsu K, Hoffmann A, Montminy M and Karin M. *Signaling pathways and genes that inhibit pathogen-induced macrophage apoptosis–CREB and NF-kappaB as key regulators.* Immunity. 2005; 23: 319–29.
Pirola L, Johnston AM and Van Obberghen E.*Modulation of insulin action.* Diabetologia. 2004; 47: 170–84.
Poulain-Godefroy O and Froguel P. *Preadipocyte response and impairment of differentiation in an inflammatory environment.* Biochem Biophys Res Commun. 2007; 356: 662–7.
Previs SF, Withers DJ, Ren JM, White MF and Shulman GI. *Contrasting effects of IRS-1 versus IRS-2 gene disruption on carbohydrate and lipid metabolism in vivo.* J Biol Chem. 2000; 275: 38990–4.
Qatanani M and Lazar MA. *Mechanisms of obesity-associated insulin resistance: many choices on the menu.* Genes Dev. 2007; 21: 1443–55.
Raetz CR. *Biochemistry of endotoxins.* Annu Rev Biochem. 1990; 59: 129–70.
Ricci R, Sumara G, Sumara I, Rozenberg I, Kurrer M, Akhmedov A, Hersberger M, Eriksson U, Eberli FR, Becher B, Boren J, Chen M, Cybulsky MI, Moore KJ, Freeman MW, Wagner EF, Matter CM and Luscher TF. *Requirement of JNK2 for scavenger receptor A-mediated foam cell formation in atherogenesis.* Science. 2004; 306: 1558–61.
Romanski SA, Nelson RM and Jensen MD. *Meal fatty acid uptake in adipose tissue: gender effects in nonobese humans.* Am J Physiol Endocrinol Metab. 2000; 279: E455–62.
Rudich A, Kanety H and Bashan N. *Adipose stress-sensing kinases: linking obesity to malfunction.* Trends Endocrinol Metab. 2007; 18: 291–9.
Schmidt AM, Hori O, Chen JX, Li JF, Crandall J, Zhang J, Cao R, Yan SD, Brett J and Stern D. *Advanced glycation endproducts interacting with their endothelial receptor induce expression of vascular cell adhesion molecule-1 (VCAM-1) in cultured human endothelial cells and in mice. A potential mechanism for the accelerated vasculopathy of diabetes.* J Clin Invest. 1995; 96: 1395–403.
Serhan CN. R*esolution phase of inflammation: novel endogenous anti-inflammatory and proresolving lipid mediators and pathways.* Annu Rev Immunol. 2007; 25: 101–37.
Sethi JK and Vidal-Puig AJ. *Thematic review series: adipocyte biology. Adipose tissue function and plasticity orchestrate nutritional adaptation.* J Lipid Res. 2007; 48: 1253–62.
Shi H, Kokoeva MV, Inouye K, Tzameli I, Yin H and Flier JS.*TLR4 links innate immunity and fatty acid-induced insulin resistance.* J Clin Invest. 2006; 116: 3015–25.
Shoelson SE, Lee J and Goldfine AB. *Inflammation and insulin resistance.* J Clin Invest. 2006; 116: 1793–801.
Strissel KJ, Stancheva Z, Miyoshi H, Perfield JW, 2nd, DeFuria J, Jick Z, Greenberg AS and Obin MS. *Adipocyte death, adipose tissue remodeling, and obesity complications.* Diabetes. 2007; 56: 2910–8.

Takahashi K, Takeya M and Sakashita N. *Multifunctional roles of macrophages in the development and progression of atherosclerosis in humans and experimental animals.* Med Electron Microsc. 2002; 35: 179–203.

Tanabe Y, Koga M, Saito M, Matsunaga Y and Nakayama K. *Inhibition of adipocyte differentiation by mechanical stretching through ERK-mediated downregulation of PPARgamma2.* J Cell Sci. 2004; 117: 3605–14.

Taskinen MR.*Diabetic dyslipidaemia: from basic research to clinical practice.* Diabetologia. 2003; 46: 733–49.

Tontonoz P, Nagy L, Alvarez JG, Thomazy VA and Evans RM. *PPARgamma promotes monocyte/macrophage differentiation and uptake of oxidized LDL.* Cell. 1998; 93: 241–52.

Tuncman G, Hirosumi J, Solinas G, Chang L, Karin M and Hotamisligil GS. *Functional in vivo interactions between JNK1 and JNK2 isoforms in obesity and insulin resistance.* Proc Natl Acad Sci USA. 2006; 103: 10741–6.

Vlassara H, Fuh H, Donnelly T and Cybulsky M. *Advanced glycation endproducts promote adhesion molecule (VCAM-1, ICAM-1) expression and atheroma formation in normal rabbits.* Mol Med. 1995; 1: 447–56.

Wei Y, Wang D, Topczewski F and Pagliassotti MJ. *Saturated fatty acids induce endoplasmic reticulum stress and apoptosis independently of ceramide in liver cells.* Am J Physiol Endocrinol Metab. 2006; 291: E275–81.

Weisberg SP, Hunter D, Huber R, Lemieux J, Slaymaker S, Vaddi K, Charo I, Leibel RL and Ferrante AW, Jr. *CCR2 modulates inflammatory and metabolic effects of high-fat feeding.* J Clin Invest. 2006; 116: 115–24.

Weisberg SP, McCann D, Desai M, Rosenbaum M, Leibel RL and Ferrante AW, Jr. *Obesity is associated with macrophage accumulation in adipose tissue.* J Clin Invest. 2003; 112: 1796–808.

Wellen KE and Hotamisligil GS. *Inflammation, stress, and diabetes.* J Clin Invest. 2005; 115: 1111–9.

Winzen R, Gowrishankar G, Bollig F, Redich N, Resch K and Holtmann H. *Distinct domains of AU-rich elements exert different functions in mRNA destabilization and stabilization by p38 mitogen-activated protein kinase or HuR.* Mol Cell Biol. 2004; 24: 4835–47.

Wolowczuk I, Verwaerde C, Viltart O, Delanoye A, Delacre M, Pot B and Grangette C. *Feeding our immune system: impact on metabolism.* Journal. 2008; doi: 10.1155/2008/639803.

Wu J and Kaufman RJ. *From acute ER stress to physiological roles of the Unfolded Protein Response.* Cell Death Differ. 2006; 13: 374–84.

Wu S, Tan M, Hu Y, Wang JL, Scheuner D and Kaufman RJ. *Ultraviolet light activates NFkappaB through translational inhibition of IkappaBalpha synthesis.* J Biol Chem. 2004; 279: 34898–902.

Wymann MP and Schneiter R. Lipid signalling in disease. *Nat Rev Mol Cell Biol.* 2008; 9: 162–76.

Xanthis A, Hatzitolios A, Koliakos G and Tatola V. *Advanced glycosylation end products and nutrition–a possible relation with diabetic atherosclerosis and how to prevent it.* J Food Sci. 2007; 72: R125–9.

Yu C, Chen Y, Cline GW, Zhang D, Zong H, Wang Y, Bergeron R, Kim JK, Cushman SW, Cooney GJ, Atcheson B, White MF, Kraegen EW and Shulman GI. *Mechanism by which fatty acids inhibit insulin activation of insulin receptor substrate-1 (IRS-1)-associated phosphatidylinositol 3-kinase activity in muscle.* J Biol Chem. 2002; 277: 50230–6.

Yuan M, Konstantopoulos N, Lee J, Hansen L, Li ZW, Karin M and Shoelson SE. *Reversal of obesity- and diet-induced insulin resistance with salicylates or targeted disruption of Ikkbeta.* Science. 2001; 293: 1673–7.

Zhao L and Ackerman SL. *Endoplasmic reticulum stress in health and disease.* Curr Opin Cell Biol. 2006; 18: 444–52.

Chapter 5
Angiogenesis in the Metabolic Syndrome

Raquel Soares

Contents

Abstract Angiogenesis is crucial both in physiological and in pathological conditions. Evidence has been gathered regarding the involvement of angiogenesis in metabolic syndrome-associated disorders, including visceral obesity, dyslipidaemias, atherosclerosis and diabetes. Indeed, metabolic syndrome is characterized by several features that actually modulate angiogenesis such as hypoxia,

R. Soares (✉)
Department of Biochemistry (U38/FCT), Faculty of Medicine, University of Porto,
Al. Prof. Hernâni Monteiro, 4200-319 Porto, Portugal
e-mail: raqsoa@med.up.pt

R. Soares, C. Costa (eds.), *Oxidative Stress, Inflammation and Angiogenesis in the Metabolic Syndrome,* DOI 10.1007/978-1-4020-9701-0_5,

inflammation, oxidative stress, hormone imbalance or hyperglycaemia. In addition, abnormal expression of angiogenic factors within disorders clustered in the so-called metabolic syndrome, have also been reported. Conversely, the vascular defects broadly observed in metabolic syndrome can be attributed to angiogenic imbalance. In the past years, efforts have been made to develop therapeutic strategies targeting angiogenesis, mostly in oncology. However, recent lines of evidence indicate that angiogenesis is required for many disorders of distinct etiopathogenic origin, namely the ones associated with metabolic syndrome. Taking into account the increasing incidence of this syndrome worldwide, as well as the absence of effective therapeutic strategies, elucidating the angiogenic pathways that play a role in these disorders may be of great interest in order to design novel therapeutic approaches against this major public health challenge.

Keywords Cardiovascular disease · Endothelial cells · Obesity · Smooth muscle cells · Type 2 diabetes mellitus

5.1 Angiogenesis

5.1.1 The Concept of Angiogenesis

Tissue vascularisation is of primordial importance both in physiological homeostasis and disease. Angiogenesis, the formation of new blood vessels from pre-existing ones, is the major contributor for physiological and pathological vascularisation. Accordingly, the terms angiogenesis, vascularisation and neovascularisation are often used in an interchangeable manner. Angiogenesis is established as the process by which an initially vascularised system is remodelled in order to provide a complex branching network of mature vasculature (Carmeliet 2005). Under physiological conditions, angiogenesis is highly coordinated by the balance between promoting and inhibiting factors, which accomplish several events such as extracellular matrix (ECM) degradation, vessel wall support cells detachment, proliferation, survival, migration of endothelial cells (EC), anastomosis, and assembling into tubular structures (Carmeliet and Jain 2000). Afterwards, a basement membrane is formed covering endothelium, regularly surrounded by mesenchymal-derived pericytes or smooth muscle cells (SMC). However, abnormal angiogenesis is frequent in many pathological situations, either in excess or insufficient magnitude, presenting often unstable and leaky vessels (Konerding et al. 2007).

5.1.2 Angiogenesis in Metabolic Syndrome (MS)

Despite deregulated neovascularisation being already reported back in the 19th century in pathology, this concept gained great impact with Judah Folkman's theory, in which angiogenesis was considered crucial for solid tumour growth and dissemination (Folkman et al. 1971). This resulted in an exponential research and development of novel knowledge regarding angiogenesis, with many different approaches being established, in order to study and target angiogenesis for therapeutic purposes. Nowadays, angiogenesis is recognized as an organizing principle, occurring associated with more than 70 distinct unrelated pathologies (Folkman 2007a, b). Accordingly, an increasing number of highly prevalent pathological situations, affecting millions of people worldwide, are angiogenic-dependent. Among these, rely disorders that constitute the MS, a cluster of several risk factors for cardiovascular disease and type 2 diabetes, namely abdominal obesity, hyperinsulinaemia, insulin resistance, impaired fasting and post-prandial glucose tolerance, arterial hypertension, dyslipidaemia, elevated free fatty acid plasma levels, microalbuminuria, hyperuricaemia, increased leptin and decreased adiponectin and plasminogen activator inhibitor type-1 (PAI-1) plasma levels (Wingard et al. 1983; Ogden et al. 2007; Rana et al. 2007; Shoelson et al. 2007 Jelski and Szmitkowski, 2007)(see Chapter 1). MS carries actually high morbidity and mortality rates, representing a major public health threat in the western world (Caterson et al. 2004).

5.2 Conditions that Stimulate Angiogenesis in MS

Together with oxidative stress and chronic inflammation (described in the previous chapters), angiogenesis constitutes another joint partner involved in the development and progression of MS. Angiogenesis can be modulated by many molecules and microenvironment features.

5.2.1 Hypoxia

Hypoxia is probably the most well-established angiogenic inducer. Hypoxic environments result in up-regulation of hypoxia inducible factor-α (HIF-α) subunits, which act as transcription factors of several angiogenic genes (Carmeliet 2000; Maxwell et al. 1999; Hirota and Semenza, 2006; Semenza et al. 2006). This is the case of vascular endothelial growth factor (VEGF), angiopoietin-2 (Ang-2) or nitric oxide synthase (NOS), whose expressions are up-regulated by HIF-α binding to responsive elements at the promotor region of these genes (Soares and Costa, 2007).

5.2.2 Chronic Inflammation

Chronic inflammation with concomitant cytokine and growth factors release is another primary stimulus for angiogenesis to occur. In fact, chronic inflammation and angiogenesis are two processes that gather together. Inflammation-induced release of angiogenic factors result in increased angiogenesis (reviewed in Costa et al. 2007). On the other hand, angiogenesis further contributes to inflammation by providing oxygen and nutrients for metabolic requirements in inflammatory sites, as well as by enabling extravasation of immune cells (Soares and Costa 2007; Costa et al. 2007). Pro-inflammatory cytokines and growth factors, secreted during inflammation, activate neighbouring EC to attract inflammatory cells, promoting their adhesion to endothelium and extravasation (Kreis and Vale 1999). The recent study by Alfranca et al. (2008), shows that prostaglandin E_2 (PGE_2) induces angiogenesis via MT1-Matrix metalloproteinase (MMP)-mediated activation of the transforming growth factor (TGF)- β/ALK5 signalling pathway.

5.2.3 Oxidative Stress

In addition, as reported in the previous chapter, reactive oxygen species (ROS) are also involved in the development and progression of vascular diseases associated with MS, such as coronary heart disease, hypertension, atherosclerosis and diabetes. Gathered evidence indicates that ROS activate intracellular signalling pathways that lead to angiogenesis (Thannickal and Fanburg 2000; Ushio-Fukai and Nakamura 2008; Polytarchou et al. 2007).

Accordingly, NAD(P)H oxidases (Nox) are a source of ROS formation from molecular oxygen (Blanchetot and Boonstra 2008; Polytarchou et al. 2007). The presence of these enzyme complexes in EC and SMC is indicative of its relevance in these cells homeostasis. Recent evidence shows that Nox up-regulate several pro-angiogenic factors, including VEGF, MMP-1, -2, and -9 (Polytarchou et al. 2007). Cyclooxygenases (COX) are also ROS-generating enzymes. The involvement of the inducible isoform COX-2 in the angiogenic process is further well recognized (Costa et al. 2002). In addition, the endothelial isoform of nitric oxide synthase (eNOS), a complex oxido-reductase, may also contribute to the generation of ROS (Vásquez-Vivar et al. 1998).

Recently, the elegant studies by Moncada and co-workers (Palacios-Callender et al. 2007; Erusalimsky and Moncada 2007), reported the involvement of NO, a well established inflammatory marker and vasodilator, on ROS production. According to this study, in certain redox conditions, NO may also be reduced by the cytochrome c oxidase enzyme complex at the mitochondrial electron transport chain (ETC). This reduction prevents oxygen from being reduced to water, leading to the accumulation of ROS, which can also be generated at the ETC. Further cross talks between Nox, eNOS, COX-2 and hypoxia in EC have been described (Bayraktutan 2004; Waypa et al. 2002; Cai et al. 2003; Fandrey et al. 1994; Kinnula et al. 1993).

5.2.4 Hormones and Growth Factors

Moreover, several hormones, including insulin, steroid hormones, leptin, ghrelin and adiponectin, also interfere with vascular remodeling and angiogenesis (Soares et al. 2003, 2004; Rocha et al. 2007; Costa et al. 2007; Tigno et al. 2003; Vona-Davis et al. 2007; Cowey and Hardy 2006). Beyond its well established role in thermogenesis, leptin has been shown to be implicated in oxidative stress and inflammation (described in the previous chapters), but has also been recognized as an important angiogenic stimulating factor in adipose tissue (AT) (Tigno et al. 2003; Frühbeck 2006). Accordingly, leptin up-regulates endothelial endothelin-1 and NOS, and enhances the expression of adhesion molecules in EC, which are essential during the angiogenic process (Frühbeck 2006). Conversely, adiponectin, whose secretion by AT is frequently decreased in patients with MS, seems to inhibit microvessel density (Vona-Davis et al. 2007). Ghrelin, another hormone associated with metabolism, obesity and appetite, increases blood flow in a dose-dependent manner (Tigno et al. 2003). Microvascular endothelial cells have been shown to express ghrelin receptor. Upon binding to its membrane receptor, ghrelin promotes EC proliferation, migration and *in vitro* angiogenesis (Li A et al. 2007).

Thus, local hypoxia, inflammation, cytokines, growth factors and the presence of imbalanced hormones characterize MS and play an effective role in angiogenesis (Li WW et al. 2007). Yet, another player in this cross-talk is the nuclear factor kappaB (NFκB), a molecule that regulates many cell fate decisions. Acting as a transcription factor for several genes, including MMPs, urokinase type of plasminogen activator (uPA), vascular and intercellular adhesion molecules (V-, I-CAM), NFκB mediates both inflammatory and angiogenic processes within the EC (Charo and Taubman, 2004; Charo and Ransohoff, 2006; Maxwell et al. 1999; Karin 2006; Karin et al. 2002).

Therefore, EC are the main targets for angiogenesis to occur. Nevertheless, the role of vessel support cells, such as SMC and pericytes, in the angiogenic process, cannot be ignored. Mural cells are the main producers of angiopoietin (Ang)-1, which binds and activates its specific receptor Tie-2 in EC, in a paracrine manner (Tsigkos and Papapetropoulus 2007). Ang-1 is able to stabilize vessels, promoting, hence, vascular integrity. In contrast, Ang-2 is mainly produced by EC, exerting an opposite effect in the vasculature. Ang-2 is frequently co-expressed with VEGF in sites of vascular remodelling, leading to effective angiogenesis. Thus, the antagonic effect of these two molecules is mandatory for vascular remodelling. But in addition to its angiogenic role, Ang-2 is also established as a pro-inflammatory molecule, leading to leukocyte recruitment and extravasation by interacting with NFκB (Fiedler et al. 2006).

Altogether, these findings emphasize the strong link between inflammation, oxidative stress and angiogenesis, three processes that are gathered together in MS.

5.3 Presence of Angiogenic Factors in MS

5.3.1 VEGF Signalling in MS

An increasing number of pro-angiogenic and anti-angiogenic factors has been reported in both physiological and pathological conditions. Yet, among all pro-angiogenic factors, VEGF represents the primary rate-limiting step of angiogenesis both in physiological and pathological conditions (Ferrara et al. 2003). By binding to its tyrosine kinase receptors (VEGFR-1 and VEGFR-2) in EC, VEGF stimulates EC survival, proliferation and migration (Wheeler-Jones et al. 1997, Gerber et al. 1998a, b). VEGF is also known as vascular permeability factor due to its marked action upon vessel permeability. In addition, VEGF is also involved in the recruitment of endothelial precursor cells (EPCs) from the bone marrow to peripheral circulation, and differentiation of these into EC in angiogenic sites (Asahara et al. 1999; reviewed in Chapter 6).

Despite the broad effects of VEGF within the angiogenic process, several other angiogenic factors have been reported: placental growth factor (PlGF, mainly present in pathological angiogenesis), acidic and basic fibrobast growth factors (a/bFGF), hepatocyte growth factor (HGF), angiopoietins, platelet-derived growth factor (PDGF), TGFβ, among many more (Yancopoulos et al. 2000; Presta et al. 2005; Miyazono et al. 1991; Ohnishi and Daikuhara 2003).

5.3.2 The Role of VLDL and its Receptor in Angiogenesis

Interestingly, the receptor of very low density lipoproteins (VLDLR) is a relevant mediator in MS (reviewed in Costa et al. 2007). Besides binding to Apo E and lipoprotein lipase, VLDLR also binds to thrombospondin (TSP)-1, uPA/ PAI-1 complex, protein/serpin complexes and tissue factor pathway inhibitor (Mikhailenko et al. 1997; Argraves et al. 1995; Kasza et al. 1997; Hembrough et al. 2004), implying a possible role of this receptor in angiogenesis as well.

5.3.3 uPA and PAI-1: A Central Role in MS Angiogenesis

Tissue type plasminogen activator (tPA) and uPA are both involved in activation of plasminogen into plasmin, which is responsible for the degradation of fibrin into soluble degradation products, as well as in cell proliferation and adhesion (Zorio et al. 2008). The activity of tPA and uPA is, thus, mediated by PAI, being PAI-1 the primordial fibrinolytic inhibitor. PAI-1 is expressed in EC, fibroblasts, SMC, adipocytes, and many epithelial cells (Beaulieu et al. 2007). Increased circulating levels of PAI-1 have been found in MS, which can be due to the fact that several features present in MS, such as adiposity or insulin resistance, enhance PAI-1. PAI-1 circulating levels have been reported to correlate with visceral adiposity and myocardial infarction (You et al. 2008). Moreover, MS features together with the presence of specific polymorphisms of the promotor region of PAI-1 are established modulators of the molecule (Zorio et al. 2008). Controversial findings have been reported regarding the angiogenic role of PAI-1. PAI-1$^{-/-}$ mice exhibit insufficient angiogenesis (McMahon et al. 2001; Masson et al. 2002). Delivery of physiological concentration of PAI-1 to these mice restores angiogenesis. However, above physiological dose, PAI-1 treatment reduces angiogenesis. Interaction of PAI-1 with ECM is likely to cause angiogenesis. In agreement, PAI-1 binds to vitronectin, preventing thus its binding to integrins, resulting, therefore, in EC migration (Ouchi et al. 2003). According to its role, increased levels of PAI-1 would down-regulate uPA and tPA activity, leading to better prognosis. However, in several MS-associated cancers, including breast cancer, PAI-1 expression is associated with poor prognosis (Beaulieu et al. 2007). This condition is known as the "PAI-1 paradox".

5.3.4 Evidence for the Presence of Angiogenic Inhibitors in MS

Apart from its angiogenic stimulating effect, there are also controversial reports concerning the up-regulation of angiogenesis inhibitory factors in MS. Pigment epithelium-derived factor (PEDF) is a potent angiogenic inhibitor, which is reduced in angiogenic eye disorders. A recent study by Yamagishi et al. (2006) demonstrated significant increased serum levels of PEDF in MS patients. Apart from its anti-angiogenic role, PEDF also exerts anti-inflammatory and anti-oxidative effects, and prevents the formation of advanced glycation products, a feature frequently observed in diabetic patients (see Chapters 3 and 4 for review) (Yamagishi et al. 2004, 2005; Inagaki et al. 2003). Whether the increase of PEDF is cause or consequence of MS remains to be elucidated. Nevertheless, it has been hypothesized that PEDF may be counter-acting the oxidative, inflammatory and angiogenic phenomena observed in MS patients.

5.4 Angiogenesis in Type 2 Diabetes Mellitus (T2DM)

5.4.1 Increased Angiogenesis in Retinopathy and Nephropathy Complications

Given its increasing incidence worldwide, diabetes will be the leading cause of mortality and morbidity in the near future. Diabetes mellitus represent one of the more prevalent pathological situations that exhibit impaired angiogenesis (Martin et al. 2003). For instance, diabetic retinopathy and nephropathy are characterized by excessive angiogenesis. Conversely, these patients often present long-term complications, such as impaired wound healing and impaired development of coronary collaterals which correspond to insufficient angiogenesis (Costa et al. 2007; Martin et al. 2003). Remarkably, diabetic patients present reduced circulating EPC (see Chapter 6). Upon activation by VEGF, vessels become leaky and angiogenic presenting decreased number of support cells. In the retina, this results in local inflammation and vascular sprouting (Adamis et al. 1999). These new vessels easily disrupt, leading to vitreous haemorrhage and retinal detachment. Similarly, in atheroma plaques proliferation of the arterial wall can also result in sprouting, plaque destabilization and eventually thrombosis (Costa et al. 2007). Therefore, absence of VEGF signalling in EC prevents arteriogenic response as well, resulting in common vessel regression.

5.4.2 Deficient Angiogenesis in Wound Healing and Coronary Heart Disease

Diabetic wound healing, a process highly dependent on angiogenesis, is of major concern due to its high prevalence among diabetic patients, as well as to the high morbidity, and aggressive and expensive intervention requirements for the treatment of these patients. Diabetic wounds present deficient angiogenic growth factors and receptors, and are further unable to recruit EPC to granuloma sites, as previously described (see Chapter 6). In addition, proteases and other molecules able to degrade ECM are intensely released in diabetic ulcers, either by cells present in the granuloma tissue or by infecting agents, such as pathogenic bacteria that frequently colonize these tissues (Li WW et al. 2007).

Hyperglycaemia, a common feature in diabetes, results frequently in glycation of many circulating factors, such as FGF, HGF, PlGF and PDGF (Duraisamy et al. 2001), implying that many other angiogenic factors may also be involved in this complex picture. The identification of these factors and their signalling pathways is essential for the development of novel therapeutic strategies.

Cardiovascular diseases include coronary heart disease, peripheral vascular disease and stroke (Jelski and Szmitkowski 2008). Coronary artery disease is significantly prevalent among diabetic patients (Bourassa and Berry 2008; Irons et al. 2006). An increase in the expression of VEGF and its two specific receptors (VEGFR-1 and VEGFR-2) was found in coronary heart disease (CHD) patients in comparison with control subjects (Sasso et al. 2005). However, in diabetic patients, despite significantly higher levels of VEGF being also reported, the expression of the two VEGFR isoforms in the myocardium was reduced when compared to non-diabetic CHD patients. In agreement with this, an effective decrease in the phosphorylated (active) form of VEGFR-2, as well as a decrease in its downstream effectors Akt and eNOS proteins have been observed (Sasso et al. 2005). These findings explain the insufficient angiogenesis observed in diabetic patients developing ischemic cardiomyopathy. Interestingly, Chou et al demonstrated that the expression of VEGF in diabetic rats was significantly decreased in myocardium of nonischemic rats compared to ischemic ones (Chou et al. 2002).

Early attempts have been made in order to design therapeutic agents to myocardial or limb ischemia (Weninger et al. 1996; Mullerat et al. 2003; Smith-McCune and Weidner 1994). In spite of the several attempts, Phase II clinical trials using FGF or VEGF gene or protein transfer had no special impact (Lekas et al. 2006; Simons and Ware 2003). The designing of angiogenic stimulatory agents to treat CHD in diabetic patients must, though, be carefully attempted, since nephropathy and retinopathy are two well established complications of diabetes that exhibit excessive angiogenesis.

5.5 Angiogenesis in Obesity

Visceral obesity constitutes a central feature in MS. The fact that obesity incidence increased more than 70% worldwide in the last twenty years (Whelton et al. 2007), together with its strong association with T2DM and cardiovascular disease, rendered obesity an emerging public health problem. Angiogenesis may mediate AT plasticity as highlighted by Folkman a few years ago (Dallabrida et al. 2003). Indeed, increase in AT mass is accompanied by expansion of capillary beds, which provide the metabolic requirements for AT remodelling. In agreement, angiogenesis is taken as a dynamic and active process in AT (Rupnick et al. 2002). Angiogenesis in AT is probably mediated by local factors, primarily released by the adipocytes (Costa et al. 2007). This has been quite well established ever since besides a fatty acid storage deposit, adipocytes were known to release several hormones and adipokines, a variety of signalling proteins implicated in regulating a huge number of processes, such as appetite, energy balance, insulin resistance/sensitivity, lipid metabolism, cell proliferation and differentiation, inflammation and angiogenesis (Cowey and Hardy 2006). That is the case of Ang1, an angiogenic factor implicated in vessel maturation, which has been shown to be expressed by adipocytes (Dallabrida et al. 2003). Inversely, its antagonist, Ang2, as well as the angiopoietin specific receptor Tie2, are expressed by adipose EC (Dallabrida et al. 2003).

Increased levels of adipokines exerting pro-angiogenic activity have been described in overweight and obese patients (Rose et al. 2004). Abrogation of VEGF signalling pathway led to decreased angiogenesis and adipose tissue mass loss (Fukumura et al. 2003). Furthermore, adipose-derived stromal cells possess angiogenic activity as well (Rehman et al. 2004; Stamm et al. 2007; Stewart et al. 2006; Sumi et al. 2007).

Besides VEGF, also leptin, Ang1, Ang2 and PlGF mediate angiogenesis in AT. For instance, Dallabrida et al. (2003) demonstrated that Ang1 expression was reduced during diet restriction both in obese and lean (wild type) mice. Leptin treatment resulted in overexpression of Ang1, confirming that adipose homeostasis, which is accomplished by this hormone, further mediates angiogenic factor. Interestingly, when Ang1 containing plasmids were injected in obese (ob/ob) mice, a significant decrease in weight gain was observed (Dallabrida et al. 2003), implying that Ang1-enhanced vascular maturation may determine AT mass. This is further emphasized by the fact that all these factors are stimulated by hypoxia and inflammation (Hausman and Richardson 2004), two mandatory conditions present in obesity.

5.6 Therapeutic Approaches Targeting Angiogenesis in MS. An Overview

Several lines of evidence indicate that angiogenesis is an active process in the development and progression of MS-associated disorders. Therefore, angiogenesis-based therapy is an unifying approach to many pathological situations that are gathered together through MS, including obesity, diabetes and CVD. Only a few agents have been designed as valuable anti-angiogenic agents. Most of these are exclusively used in oncology. Given the recent establishment of the angiogenic process as an organizing principle among several pathological situations, new attempts are now being made in order to spread these anti-angiogenic agents towards other disorders presenting excessive angiogenesis as well. Many other factors and signalling pathways implicated in pathological angiogenesis are under intense research to develop new strategies able to directly or indirectly target angiogenesis. These attempts include advances in targeting distinct angiogenic factors, delivery improvement of pharmacological agents, combination therapy, cell-based therapy, and identification and development of new exogenous compounds able to affect angiogenesis, as is the case of polyphenolic compounds (see Chapter 7).

Treatment of obesity has been restricted to a few agents that target mainly brain receptors, inhibiting appetite and energy balance signalling pathways (Leonhardt et al. 1999; Bray 1999). However, those agents display significant side-effects. Pre-clinical studies using anti-angiogenic molecules such as prohibitin seem to result in obesity withdrawal (Kolonin et al. 2004). Therefore, elucidating putative new vascular targets may lead to novel approach to reduce AT mass.

Emerging advances shed light into the involvement of angiogenesis in diabetes. Angiogenesis is decreased in chronic and delayed wound healing, a major problem in diabetic patients. As reported by William Li et al. (2007), there are actually five major categories of wound interventions that lead to angiogenesis stimulation: growth factor therapies; tissue engineering products; bioactive matrices; mechanical systems; and hyperbaric oxygen therapy. Conversely, diabetic retinopathy involves the up-regulation of many angiogenic and inflammatory factors including VEGF, HIFα, adhesion molecules, and TNFα, which end up in strong neovascularization. Nowadays several anti-angiogenic agents are being used in diabetic retinopathy treatment with promising results in visual acuity of these patients (Costa et al. 2007; Soares and Costa 2007; Vinores 2007). However, given the complexity of diabetes, systemic stimulation of angiogenesis for tissues presenting insufficient vascularization may result in increased angiogenesis in tissues where angiogenesis is already in excess, namely retina or kidney. Furthermore, delivery of systemic pro-angiogenic factors, such as VEGF or FGF, may lead to hypotension and edema, anaemia, thrombocytopenia and renal toxicity (Luttun et al. 2002, Martin et al. 2003).

In conclusion, although several MS-associated conditions are angiogenesis-related, many questions regarding the development of therapeutic approaches addressing angiogenesis still remain. Basic and translational research is lacking in order to validate angiogenesis-based therapeutic strategies, to evaluate toxicity, or to predict which patients may benefit from these therapies.

References

Adamis AP, Aiello LP, D'Amato RA. *Angiogenesis and ophthalmic disease.* Angiogenesis. 1999; 3: 9–14.

Alfranca A, López-Oliva JM, Genís L, López-Maderuelo D, Mirones I, Salvado D, Quesada AJ, Arroyo AG, Redondo JM. *PGE2 induces angiogenesis via MT1-MMP-mediated activation of the TGFbeta/Alk5 signaling pathway.* Blood. 2008; 112: 1120–8.

Argraves KM, Battey FD, MacCalman CD, McCrae KR, Gåfvels M, Kozarsky KF, Chappell DA, Strauss JF 3rd, Strickland DK. *The very low density lipoprotein receptor mediates the cellular catabolism of lipoprotein lipase and urokinase-plasminogen activator inhibitor type I complexes.* J Biol Chem. 1995; 270: 26550–7.

Asahara T, Takahashi T, Masuda H, Kalka C, Chen D, Iwaguro H, Inai Y, Silver M, Isner JM. *VEGF contributes to postnatal neovascularization by mobilizing bone marrow-derived endothelial progenitor cells.* EMBO J. 1999; 18: 3964–72.

Bayraktutan U. *Nitric oxide synthase and NAD(P)H oxidase modulate coronary endothelial cell growth.* J Mol Cell Cardiol. 2004; 36: 277–86.

Beaulieu LM, Whitley BR, Wiesner TF, Rehault SM, Palmieri D, Elkahloun AG, Church FC. *Breast cancer and metabolic syndrome linked through the plasminogen activator inhibitor-1 cycle.* Bioessays. 2007; 29: 1029–38.

Blanchetot C, Boonstra J. *The ROS-NOX connection in cancer and angiogenesis.* Crit Rev Eukaryot Gene Expr. 2008; 18: 35–45.

Bourassa MG, Berry C. *Prevention and noninvasive management of coronary atherosclerosis in patients with diabetes.* Curr Atheroscler Rep. 2008; 10: 106–16.

Bray GA. *Uses and misuses of the new pharmacotherapy of obesity.* Ann Med. 1999; 31: 1–3.

Cai H, Griendling KK, Harrison DG. *The vascular NAD(P)H oxidases as therapeutic targets in cardiovascular diseases.* Trends Pharmacol Sci. 2003; 24: 471–8.

Carmeliet P, Jain RK. *Angiogenesis in cancer and other diseases.* Nature. 2000; 407: 249.

Carmeliet P. *Angiogenesis in life, disease and medicine.* Nature. 2005; 438: 932–6.

Carmeliet P. *VEGF gene therapy: stimulating angiogenesis or angioma-genesis?* Nature Med. 2000; 6: 1002–3.

Caterson ID, Hubbard V, Bray GA, Grunstein R, Hansen BC, Hong Y, et al. *Prevention Conference VII: Obesity, a worldwide Epidemic related to heart disease and stroke: Group III: Worldwide Comorbidities of Obesity.* Circulation. 2004; 110: 476–83.

Charo IF, Ransohoff RM. *The many roles of chemokines and chemokine receptors in inflammation.* N Engl J Med. 2006; 354: 610–21.

Charo IF, Taubman MB. *Chemokines in the pathogenesis of vascular disease.* Circ Res. 2004; 95: 858–66.

Chou E, Suzuma I, Way KJ, Opland D, Clermont AC, Naruse K, Suzuma K, Bowling NL, Vlahos CJ, Aiello LP, King GL. *Decreased cardiac expression of vascular endothelial growth factor and its receptors in insulin-resistant and diabetic States: a possible explanation for impaired collateral formation in cardiac tissue.* Circulation. 200; 105: 373–9.

Costa C, Incio J, Soares R. *Angiogenesis and chronic inflammation: cause or consequence?* Angiogenesis. 2007; 10: 149–66.

Costa C, Soares R, Reis-Filho JS, Leitão D, Amendoeira I, Schmitt FC. *Cyclo-oxygenase 2 expression is associated with angiogenesis and lymph node metastasis in human breast cancer.* J Clin Pathol. 2002; 55: 429–34.

Cowey S, Hardy RW. *The metabolic syndrome: A high-risk state for cancer?* Am J Pathol. 2006; 69: 1505–22.

Dallabrida SM, Zurakowski D, Shih SC, Smith LE, Folkman J, Moulton KS, Rupnick MA. *Adipose tissue growth and regression are regulated by angiopoietin-1.* Biochem Biophys Res Commun. 2003; 311: 563–71.

Duraisamy Y, Slevin M, Smith N, Bailey J, Zweit J, Smith C, Ahmed N, Gaffney J. *Effect of glycation on basic fibroblast growth factor induced angiogenesis and activation of associated signal transduction pathways in vascular endothelial cells: possible relevance to wound healing in diabetes.* Angiogenesis. 2001; 4: 277–88.

Erusalimsky JD, Moncada S. *Nitric oxide and mitochondrial signaling: from physiology to pathophysiology.* Arterioscler Thromb Vasc Biol. 2007; 27: 2524–31.

Fandrey J, Frede S, Jelkmann W. *Role of hydrogen peroxide in hypoxia-induced erythropoietin production.* Biochem J. 1994; 303: 507–10.

Ferrara N, Gerber HP, LeCouter J. *The biology of VEGF and its receptors.* Nat Med. 2003; 9: 669–76.

Fiedler U, Reiss Y, Scharpfenecker M, Grunow V, Koidl S, Thurston G, Gale NW, Witzenrath M, Rosseau S, Suttorp N, Sobke A, Herrmann M, Preissner KT, Vajkoczy P, Augustin HG. *Angiopoietin-2 sensitizes endothelial cells to TNF-alpha and has a crucial role in the induction of inflammation.* Nat Med. 2006; 12: 235–9.

Folkman J. *Angiogenesis: an organizing principle for drug discovery?* Nat Rev Drug Discov. 2007a; 6: 273–86.

Folkman J. *Is angiogenesis an organizing principle in biology and medicine?* J Pediatr Surg. 2007b; 42: 1–11.

Folkman J. *Tumor angiogenesis: therapeutic implications.* N Engl J Med. 1971; 285: 1182–6.

Frühbeck G. *Intracellular signalling pathways activated by leptin.* Biochem J. 2006; 393: 7–20.

Fukumura D, Ushiyama A, Duda DG, Xu L, Tam J, Krishna V, Chatterjee K, Garkavtsev I, Jain RK. *Paracrine regulation of angiogenesis and adipocyte differentiation during in vivo adipogenesis.* Circ Res. 2003; 93: e88–97.

Gerber HP, Dixit V, Ferrara N. *Vascular endothelial growth factor induces expression of the antiapoptotic proteins Bcl-2 and A1 in vascular endothelial cells.* J Biol Chem. 1998a; 273: 13313–6.

Gerber HP, McMurtrey A, Kowalski J, Yan M, Keyt BA, Dixit V, Ferrara N. *Vascular endothelial growth factor regulates endothelial cell survival through the phosphatidylinositol 3′-kinase/Akt signal transduction pathway. Requirement for Flk-1/KDR activation.* J Biol Chem. 1998b; 273: 30336–43.

Hausman GJ, Richardson RL. *Adipose tissue angiogenesis.* J Anim Sci. 2004; 82: 925–34.

Hembrough TA, Ruiz JF, Swerdlow BM, Swartz GM, Hammers HJ, Zhang L, Plum SM, Williams MS, Strickland DK, Pribluda VS. *Identification and characterization of a very low density lipoprotein receptor-binding peptide from tissue factor pathway inhibitor that has antitumor and antiangiogenic activity.* Blood. 2004; 103: 3374–80.

Hirota K, Semenza GL. *Regulation of angiogenesis by hypoxia-inducible factor 1.* Crit Rev Oncol Hematol. 2006; 59: 15–26.

Inagaki Y, Yamagishi S, Okamoto T, Takeuchi M, Amano S. *Pigment epithelium-derived factor prevents advanced glycation end products-induced monocyte chemoattractant protein-1 production in microvascular endothelial cells by suppressing intracellular reactive oxygen species generation.* Diabetologia. 2003; 46: 284–287.

Irons BK, Greene RS, Mazzolini TA, Edwards KL, Sleeper RB. *Implications of rosiglitazone and pioglitazone on cardiovascular risk in patients with type 2 diabetes mellitus.* Pharmacotherapy. 2006; 26: 168–81.

Jelski W, Szmitkowski M. *Effect of ethanol on metabolic syndrome.* Pol Merkur Lekarski. 2007; 24: 131–3.

Karin M, Cao Y, Greten FR, Li ZW. *NF-kappaB in cancer: from innocent bystander to major culprit.* Nat Rev Cancer. 2002; 2: 301–10.

Karin M. *Nuclear factor-kappaB in cancer development and progression.* Nature. 2006; 441: 431–6.

Kasza A, Petersen HH, Heegaard CW, Oka K, Christensen A, Dubin A, Chan L, Andreasen PA. *Specificity of serine proteinase/serpin complex binding to very-low-density lipoprotein receptor and alpha2-macroglobulin receptor/low-density-lipoprotein-receptor-related protein.* Eur J Biochem. 1997; 248: 270–81.

Kinnula VL, Mirza Z, Crapo JD, Whorton AR. *Modulation of hydrogen peroxide release from vascular endothelial cells by oxygen.* Am J Respir Cell Mol Biol. 1993; 9: 603–9.

Kolonin MG, Saha PK, Chan L, Pasqualini R, Arap W. *Reversal of obesity by targeted ablation of adipose tissue.* Nat Med. 2004; 10: 625–32.

Konerding M, Ravnic D, Wolloscheck T. *Heterogeneity of blood vessels: physiological versus pathological angiogenesis.* In: Maragoudakis ME; Papadimitriou E (ed.) Angiogenesis. Basic science and clinical applications, 1st edn. Transworld Research Network. 2007; 1–16.

Kreis T, Vale R (eds). *Guidebook to the extracellular matrix, anchor, and adhesion proteins*. Oxford University press, UK. 1999.

Lekas M, Lekas P, Latter DA, Kutryk MB, Stewart DJ. *Growth factor-induced therapeutic neovascularization for ischaemic vascular disease: time for a re-evaluation?* Curr Opin Cardiol. 2006; 21: 376–84.

Leonhardt M, Hrupka B, Langhans W. *New approaches in the pharmacological treatment of obesity*. Eur J Nutr. 1999; 38: 1–13.

Li A, Cheng G, Zhu GH, Tarnawski AS. *Ghrelin stimulates angiogenesis in human microvascular endothelial cells: Implications beyond GH release*. Biochem Biophys Res Commun. 2007; 353: 238–43.

Li WW, Hutnik M, Li V. *Angiogenesis-based medicine: principles and practice for disease prevention and intervention*. In: Maragoudakis ME; Papadimitriou E (ed.) Angiogenesis. Basic science and clinical applications, 1st edn. Transworld Research Network. 2007; 377–417.

Luttun A, Tjwa M, Carmeliet P. *Placental growth factor (PlGF) and its receptor Flt-1 (VEGFR-1): novel therapeutic targets for angiogenic disorders*. Ann N Y Acad Sci. 2002; 979: 80–93.

Martin A, Komada MR, Sane DC. *Abnormal angiogenesis in diabetes mellitus*. Med Res Rev. 2003; 23: 117–45.

Masson V, Devy L, Grignet-Debrus C, Bernt S, Bajou K, Blacher S, Roland G, Chang Y, Fong T, Carmeliet P, Foidart JM, Noël A. *Mouse aortic ring assay: a new approach of the molecular genetics of angiogenesis*. Biol Proced Online. 2002; 4: 24–31.

Maxwell PH, Wiesener MS, Chang GW, Clifford SC, Vaux EC, Cockman ME, Wykoff CC, Pugh CW, Maher ER, Ratcliffe PJ. *The tumour suppressor protein VHL targets hypoxia-inducible factors for oxygen-dependent proteolysis*. Nature. 1999; 399: 271–5.

McMahon GA, Petitclerc E, Stefansson S, Smith E, Wong MK, Westrick RJ, Ginsburg D, Brooks PC, Lawrence DA. *Plasminogen activator inhibitor-1 regulates tumor growth and angiogenesis*. J Biol Chem. 2001; 276: 33964–8.

Mikhailenko I, Krylov D, Argraves KM, Roberts DD, Liau G, Strickland DK.*Cellular internalization and degradation of thrombospondin-1 is mediated by the amino-terminal heparin binding domain (HBD). High affinity interaction of dimeric HBD with the low density lipoprotein receptor-related protein*. J Biol Chem. 1997; 272: 6784–91.

Miyazono K, Usuki K, Heldin CH. *Platelet-derived endothelial cell growth factor*. Prog Growth Factor Res. 1991; 3: 207–17.

Mullerat J, Wong Te Fong LF, Davies SE, Winslet MC, Perrett CW. *Angiogenesis in anal warts, anal intraepithelial neoplasia and anal squamous cell carcinoma*. Colorectal Dis. 2003; 5: 353–7.

Ogden CL, Yanovski SZ, Carol MD, Flegal KM. *The epidemiology of obesity*. Gastroenterology. 2007; 132: 2087–2102.

Ohnishi T, Daikuhara Y. *Hepatocyte growth factor/scatter factor in development, inflammation and carcinogenesis: its expression and role in oral tissues*. Arch Oral Biol. 2003; 48: 797–804.

Ouchi N, Kihara S, Funahashi T, Nakamura T, Nishida M, Kumada M, Okamoto Y, Ohashi K, Nagaretani H, Kishida K, Nishizawa H, Maeda N, Kobayashi H, Hiraoka H, Matsuzawa Y. *Reciprocal association of C-reactive protein with adiponectin in blood stream and adipose tissue*. Circulation. 2003; 107: 671–4.

Palacios-Callender M, Hollis V, Mitchison M, Frakich N, Unitt D, Moncada S. *Cytochrome c oxidase regulates endogenous nitric oxide availability in respiring cells: a possible explanation for hypoxic vasodilation*. Proc Natl Acad Sci USA. 2007; 104: 18508–13.

Polytarchou C, Hatziapostolou M, Poimenidi E, Papadimitriou E. *Reactive oxygen species as mediators of angiogenesis signaling*. In: Maragoudakis ME, Papadimitriou E (ed.) Angiogenesis. Basic science and clinical applications, 1st edn. Transworld Research Network. 2007; 207–27.

Presta M, Dell'Era P, Mitola S, Moroni E, Ronca R, Rusnati M. *Fibroblast growth factor/fibroblast growth factor receptor system in angiogenesis*. Cytokine Growth Factor Rev. 2005; 16: 159–78.

Rana J, Nieuwdorp M, Jukema J, Kastelein J. *Cardiovascular metabolic syndrome – an interplay of obesity, inflammation, diabetes and coronary heart disease*. Diabetes Obes Metab. 2007; 9: 329–32.

Rehman J, Traktuev D, Li J, Merfeld-Clauss S, Temm-Grove CJ, Bovenkerk JE, Pell CL, Johnstone BH, Considine RV, March KL. *Secretion of angiogenic and antiapoptotic factors by human adipose stromal cells.* Circulation. 2004; 109: 1292–8.

Rocha A, Azevedo I, Soares R. *Anti-angiogenic effects of imatinib target smooth muscle cells but not endothelial cells.* Angiogenesis. 2007; 10: 279–86.

Rose DP, Komninou D, Stephenson GD. *Obesity, adipocytokines, and insulin resistance in breast cancer.* Obes Rev. 2004; 5: 153–65.

Rupnick MA, Panigrahy D, Zhang CY, Dallabrida SM, Lowell BB, Langer R, Folkman MJ. *Adipose tissue mass can be regulated through the vasculature.* Proc Natl Acad Sci USA. 2002; 99: 10730–5.

Sasso FC, Torella D, Carbonara O, Ellison GM, Torella M, Scardone M, Marra C, Nasti R, Marfella R, Cozzolino D, Indolfi C, Cotrufo M, Torella R, Salvatore T. *Increased vascular endothelial growth factor expression but impaired vascular endothelial growth factor receptor signaling in the myocardium of type 2 diabetic patients with chronic coronary heart disease.* J Am Coll Cardiol. 2005; 46: 827–34.

Semenza GL, Shimoda LA, Prabhakar NR. *Regulation of gene expression by HIF-1.* Novartis Found Symp. 2006; 272: 2–8.

Shoelson SE, Herrero L, Naaz A. *Obesity, inflammation and insulin resistance.* Gastroenterology 2007; 132: 2169–2180.

Simons M, Ware JA. Therapeutic angiogenesis in cardiovascular disease. Nat Rev Drug Discov. 2003; 2: 863–71.

Smith-McCune KK, Weidner N. *Demonstration and characterization of the angiogenic properties of cervical dysplasia.* Cancer Res. 1994; 54: 800–4.

Soares R, Balogh G, Guo S, Gärtner F, Russo J, Schmitt F. *Evidence for the notch signaling pathway on the role of estrogen in angiogenesis.* Mol Endocrinol. 2004; 18: 2333–43.

Soares R, Costa C. *Angiogenesis and inflammatory diseases: current concepts and therapeutic perspectives.* In: Maragoudakis ME; Papadimitriou E (ed.) Angiogenesis. Basic science and clinical applications, 1st edn. Transworld Research Network. 2007; 511–47.

Soares R, Guo S, Gärtner F, Schmitt FC, Russo J. *17 beta -estradiol-mediated vessel assembly and stabilization in tumor angiogenesis requires TGF beta and EGFR crosstalk.* Angiogenesis. 2003; 6: 271–81.

Stamm C, Kleine HD, Choi YH, Dunkelmann S, Lauffs JA, Lorenzen B, David A, Liebold A, Nienaber C, Zurakowski D, Freund M, Steinhoff G. *Intramyocardial delivery of CD133+ bone marrow cells and coronary artery bypass grafting for chronic ischemic heart disease: safety and efficacy studies.* J Thorac Cardiovasc Surg. 2007; 133: 717–25.

Stewart DJ, Hilton JD, Arnold JM, Gregoire J, Rivard A, Archer SL, Charbonneau F, Cohen E, Curtis M, Buller CE, Mendelsohn FO, Dib N, Page P, Ducas J, Plante S, Sullivan J, Macko J, Rasmussen C, Kessler PD, Rasmussen HS. *Angiogenic gene therapy in patients with nonrevascularizable ischemic heart disease: a phase 2 randomized, controlled trial of AdVEGF(121) (AdVEGF121) versus maximum medical treatment.* Gene Ther. 2006; 13: 1503–11.

Sumi M, Sata M, Toya N, Yanaga K, Ohki T, Nagai R. *Transplantation of adipose stromal cells, but not mature adipocytes, augments ischemia-induced angiogenesis.* Life Sci. 2007; 80: 559–65.

Thannickal V, Fanburg BL. *Reactive oxygen species in cell signaling.* Am J Physiol Lung Cell Mol Physiol. 2000; 279: L1005–28.

Tigno XT, Selaru IK, Angeloni SV, Hansen BC. *Is microvascular flow rate related to ghrelin, leptin and adiponectin levels?* Clin Hemorheol Microcirc. 2003; 29: 409–16.

Tsigkos S, Papapetropoulus A. *The angiopoietins: linking angiogenesis and inflammation.* In: Maragoudakis ME; Papadimitriou E (ed.) Angiogenesis. Basic science and clinical applications, 1st edn. Transworld Research Network. 2007; 79–93.

Ushio-Fukai M, Nakamura Y. *Reactive oxygen species and angiogenesis: NADPH oxidase as target for cancer therapy.* Cancer Lett. 2008; 266: 37–52.

Vásquez-Vivar J, Kalyanaraman B, Martásek P, Hogg N, Masters BS, Karoui H, Tordo P, Pritchard KA Jr. *Superoxide generation by endothelial nitric oxide synthase: the influence of cofactors.* Proc Natl Acad Sci U S A. 1998; 95: 9220–5.

Vinores S *Anti-VEGF therapy for ocular vascular diseases.* In: Maragoudakis ME; Papadimitriou E (ed.) Angiogenesis. Basic science and clinical applications, 1st edn. Transworld Research Network. 2007; 467–81.

Vona-Davis L, Howard-McNatt M, Rose DP. *Adiposity, type 2 diabetes and the metabolic syndrome in breast cancer.* Obes Rev. 2007; 8: 395–408.

Waypa GB, Marks JD, Mack MM, Boriboun C, Mungai PT, Schumacker PT. *Mitochondrial reactive oxygen species trigger calcium increases during hypoxia in pulmonary arterial myocytes.* Circ Res. 2002; 91: 719–26.

Weninger W, Uthman A, Pammer J, Pichler A, Ballaun C, Lang IM, Plettenberg A, Bankl HC, Stürzl M, Tschachler E. *Vascular endothelial growth factor production in normal epidermis and in benign and malignant epithelial skin tumors.* Lab Invest. 1996; 75: 647–57.

Wheeler-Jones C, Abu-Ghazaleh R, Cospedal R, Houliston RA, Martin J, Zachary I. *Vascular endothelial growth factor stimulates prostacyclin production and activation of cytosolic phospholipase A2 in endothelial cells via p42/p44 mitogen-activated protein kinase.* FEBS Lett. 1997; 420: 28–32.

Whelton H, Harrington J, Crowley E, Kelleher V, Cronin M, Perry IJ. *Prevalence of overweight and obesity on the island of Ireland: results from the North South Survey of Children's Height, Weight and Body Mass Index, 2002.* BMC Public Health. 2007; 7: 187.

Wingard D, Barret-Conner E, Criqui M, Suarez L: *Clustering of heart disease risk factors in diabetic compared to non-diabetic adults.* Am J Epidemiol. 1983; 117: 19–26.

Yamagishi S, Adachi H, Abe A, Yashiro T, Enomoto M, Furuki K, Hino A, Jinnouchi Y, Takenaka K, Matsui T, Nakamura K, Imaizumi T. *Elevated serum levels of pigment epithelium-derived factor in the metabolic syndrome.* J Clin Endocrinol Metab. 2006; 91: 2447–50.

Yamagishi S, Inagaki Y, Nakamura K, Abe R, Shimizu T, Yoshimura A, Imaizumi T. *Pigment epithelium-derived factor inhibits TNF-α-induced interleukin-6 expression in endothelial cells by suppressing NADPH oxidase-mediated reactive oxygen species generation.* J Mol Cell Cardiol. 2004; 37: 497–506.

Yamagishi S, Nakamura K, Ueda S, Kato S, Imaizumi T. *Pigment epithelium-derived factor (PEDF) blocks angiotensin II signaling in endothelial cells via suppression of NADPH oxidase: a novel anti-oxidative mechanism of PEDF.* Cell Tissue Res. 2005; 320: 437–445.

Yancopoulos GD, Davis S, Gale NW, Rudge JS, Wiegand SJ, Holash J. *Vascular-specific growth factors and blood vessel formation.* Nature. 2000; 407: 242–8.

You T, Nicklas BJ, Ding J, Penninx BW, Goodpaster BH, Bauer DC, Tylavsky FA, Harris TB, Kritchevsky SB. *The metabolic syndrome is associated with circulating adipokines in older adults across a wide range of adiposity.* J Gerontol A Biol Sci Med Sci. 2008; 63: 414–9.

Zorio E, Gilabert-Estellés J, España F, Ramón LA, Cosín R, Estellés A. *Fibrinolysis: the key to new pathogenetic mechanisms.* Curr Med Chem. 2008; 15: 923–9.

Chapter 6
Role of Endothelial Progenitor Cells in the Metabolic Syndrome

Carla Costa

Contents

Abstract The discovery of postnatal vasculogenesis and of the important roles played by Endothelial Progenitor Cells (EPCs) was a landmark in vascular biology that forever has changed the concept of neovascularization. In Metabolic Syndrome

---

C. Costa (✉)
Department of Biochemistry (U38/FCT) and Laboratory for Molecular Cell Biology, Faculty of Medicine, University of Porto, Al. Prof. Hernâni Monteiro, 4200-319 Porto, Portugal
e-mail: carcosta@med.up.pt

R. Soares, C. Costa (eds.), *Oxidative Stress, Inflammation and Angiogenesis in the Metabolic Syndrome,* DOI 10.1007/978-1-4020-9701-0_6,

(MS) most of EPCs biological functions seem to be impaired and associated with deficient vascular repair, with the maintenance of endothelial dysfunction conditions and the progression of atherosclerosis. The therapeutic control of MS-associated cardiovascular risk factors may restore some of EPCs abrogated functional activities preventing cardiovascular disease development.

This review summarizes current data concerning EPCs biological features in MS and provides a therapeutic outline on the beneficial effects of restoring endogenous vasculogenesis mechanisms in the MS clinical setting.

Keywords Cardiovascular risk factors · Endothelial dysfunction · Endothelial progenitor cells · Metabolic syndrome · Vasculogenesis

6.1 Endothelial Progenitor Cells (EPCs) and Postnatal Vasculogenesis – A Fallen Dogma

6.1.1 Vasculogenesis and Angiogenesis

Blood vessels constitute the first organ in the embryo and form the largest network in our body (Risau et al. 1995). Two important processes drive the development of the vascular system, vasculogenesis and angiogenesis. Vasculogenesis occurs during early embryonic development and mediates the *de novo* vessel formation from Endothelial Progenitor Cells (EPCs) or angioblasts of mesodermal origin, which differentiate in endothelial cells (ECs) assembling into a primary capillary plexus (Risau et al. 1995). After, this primitive vascular network expands by angiogenesis, where new blood vessels arise from the proliferation and migration of the pre-existing ECs (Folkman 1984; Risau 1997). During several decades it was thought that in the adult, vascular growth and remodelling was solely dependent on the activation of local angiogenesis, and that the process of vasculogenesis was restricted to embryonic development. This long-lasting belief has come to an end ten years ago.

6.1.2 Postnatal EPCs and Adult Vasculogenesis

The quest for EPCs and vasculogenesis in adult life took several decades of intensive research. In the late 1990s, Asahara and co-workers, demonstrated that a population of $CD34^+$ angioblast-like circulating EPCs could be isolated from peripheral blood of adult individuals and when cultured *in vitro* had an increased proliferation rate and exhibited endothelial morphological and functional properties (Asahara et al. 1997). Besides the reported functional characteristics of these endothelial progenitor-like cells, some questions and doubts concerning this population identity brought controversy into the field. For instance, besides the distinguished high proliferation capacity described, no other phenotypic characteristics were specific of EPCs, since the CD34 antigen used for cell isolation, is also present in subgroups of hematopoietic stem/progenitor cells and mature ECs. Nonetheless, the pioneering work of Asahara et al. was the first to propose that the process of vasculogenesis could occur in adult life. This study has marked the beginning of the end of a prevailing dogma denying the existence of postnatal vasculogenesis. Subsequent studies, have then provided clear evidence for the existence of circulating EPCs, which were mobilized from the bone marrow (BM) to the peripheral circulation, migrated and partaken *in vivo* in the development of vascular networks, by differentiating into functional, mature ECs (Shi et al. 1998; Asahara et al. 1999a; Asahara et al. 1999b; Rafii 2000; Lyden et al. 2001). In order to be mobilized from the BM and home to neovascular sites, EPCs are thought to respond to specific angiogenic stimuli. It has been reported that increased levels of chemokines, such as the Vascular Endothelial Growth Factor (VEGF) and Stromal Derived Factor (SDF)-1, may induce several cellular pathways/mechanisms which promote EPCs release, mobilization and recruitment (Asahara et al. 1999a; Aicher et al. 2005; Smythe et al. 2008). Although further studies are required to completely elucidate all the mechanisms involved, it is thought that peripheral tissue secretion of VEGF induces EPCs release from the BM microenvironment, by activating matrix metalloproteinase (MMP)-9, which by interfering with EPCs interactions with stromal cells allows them to disengage (Heissig et al. 2002). Additionally, activation of BM endothelial Nitric Oxide Synthase (eNOS) and consequent increased eNO levels, influences EPCs passage through BM sinusoidal endothelium and their entrance in the blood stream to further be recruited to neoangiogenic foci (Aicher et al. 2003; Aicher et al. 2005). Besides their differentiation into ECs upon arrival to neovascular places, EPCs may additionally contribute to neovascularization through the paracrine release of several pro-angiogenic growth factors, promoting local EC proliferation and migration, and a consequent enhancement of angiogenesis (reviewed by Hristov et al. 2003; Grunewald et al. 2006). Studies carried on experimental models have shown that postnatal vasculogenesis could take place under certain physiological and pathological settings such as, vascular homeostasis and repair (Asahara et al. 1999a; Shantsila et al. 2007), wound healing (Asahara et al. 1999b), bone regeneration (Matsumoto et al. 2008), myocardial and limb ischemia (Asahara et al. 1999a, b; Takahashi et al. 1999), and tumor neovascularisation (Asahara et al. 1999a, b; Lyden et al. 2001). Further research in adult patients, has demonstrated that EPCs recruitment and homing, may occur after vascular trauma in burn individuals (Gill et al. 2001), following acute myocardial infarction (Shintani et al. 2001) and that a percentage of BM-derived EPCs can contribute to human tumour-associated neovasculature (Peters et al. 2005). Further, it has also been proposed that other of BM-derived hematopoietic/myeloid progenitor cells, named as accessory cells, could be co-recruited to tumour neoangiogenesis foci, and support vascular growth in a paracrine fashion (Lyden et al. 2001; Kaplan et al. 2005; Grunewald et al. 2006; De Palma et al. 2005). Further studies are required to examine and confirm accessory cells putative role in additional pathophysiological situations.

6.1.3 Phenotypic and Functional Characterization of Adult EPCs

Although a great deal of interest in EPCs and in their revascularization potential has been raised since their discovery, much controversy has accompanied this research field over time. One of the main arguments concerned the phenotypic characterization of the EPC population, since their isolation and identification were mostly hampered by the lack of EPCs- specific surface markers. Some consensus has been reached and EPCs were considered to be the cell population characterized by the concomitant expression of: the early hematopoietic stem cell markers CD34 and CD133 (former AC133) and VEGF receptor-2 (VEGFR-2) (Peichev et al. 2000; reviewed by Hristov et al. 2003; Shmelkov et al. 2008). Upon differentiating, EPCs start exhibiting classical EC morphology and characteristics, such as the expression of von Willebrand factor and vascular endothelial cadherin, and the capacity to uptake acetylated low-density lipoprotein (LDL) (Shi et al. 1998; Peichev et al. 2000). Novel cumulating evidence has suggested that two distinct populations of adult EPCs displaying different phenotypes and cell surface antigens may be grown out from peripheral blood mononuclear cells. Although the biology and nature of these endothelial progenitor-like cell populations is not fully understood, they presented distinct characteristics from differentiated ECs, showing different vasculogenic features *in vitro* (Gulati et al. 2003; Hur et al. 2004). In culture, one EPC population seems to form "Early Outgrowth Colonies" (EOCs), whereas the other gives rise to "Late Outgrowth Colonies" (LOCs). EOCs and LOCs are primarily characterized based on their morphology and chronology of appearance following *in vitro* culture. EOCs appear in culture within seven days, exhibit spindle-shaped morphology emanating from a central cluster of cells, having a peak growth at two-three weeks, after which they cannot be further expanded (Gulati et al. 2003; Hur et al. 2004). LOCs generally appear after three weeks, exhibit a "classic endothelial" phenotype and have an increased expansion potential (Gulati et al. 2003; Hur et al. 2004; Ishikawa 2004; Yoder et al. 2007). LOCs seem more capable of *in vitro* morphogenesis into capillary tubes, the best approximate true definition of an EPC, a competent progenitor cell whose terminally differentiated progeny are mature ECs (Gulati et al. 2003; Hur et al. 2004; Yoder et al. 2007). This capillary-forming capacity is minimal or nonexistent within EOCs, which is thought to have a paracrine role by supporting LOCs differentiation and capillary formation, through the release of pro-angiogenic molecules and by inducing the activation of MMPs (Gulati et al. 2003; Hur et al. 2004; Yoon et al. 2005). Phenotypically, EOCs express the monocyte/macrophage marker CD14, which is absent among mature LOCs, and both populations may concomitantly express CD34, CD133 and VEGFR-2 (Gulati et al. 2003; Yoon et al. 2005). Nonetheless, most of the reports involving *in vitro* and *in vivo* EPCs studies have performed cell isolation based on the classic antigen triad, CD34, CD133, VEGFR-2, and in further cell characterization do not usually make a distinction between EOCs and LOCs.

Overall and most importantly, the identification of adult EPCs has changed the concept of neovascularization processes, being currently accepted that both local angiogenesis and systemic vasculogenesis are activated and co-orchestrate the formation of novel vasculature.

6.1.4 *The Relevance of EPCs in Clinical Settings*

An increasing body of evidence has definitely revealed the important role of EPCs on postnatal neovascularization and a growing interest into their potential use in the clinical setting has emerged. It has been suggested that EPCs could be used for cell-based therapy (Sepúlveda et al. 2007), as prognostic factor (Ho et al. 2006; Pircher et al. 2008; Naik et al. 2008), as therapeutic outcome predictor and/or as disease biomarker (reviewed by DePrimo et al. 2007; Michowitz et al. 2007; Nonaka-Sarukawa et al. 2007). In fact, due to its vasculogenic potential, it was proposed that EPCs transplantation into ischemic tissues could be an efficient treatment approach to promote vessel formation and improve blood perfusion (reviewed by Sepúlveda et al. 2007). Further, as EPCs play an important role on tumour neovascularization, it was suggested that evaluating the number of circulating $CD34^{+}VEGFR\text{-}2^{+}CD133^{+}$ EPCs in cancer patients, could be a potential prognostic factor and an indirect assessment of the neovascular potential of a growing tumour (Ho et al. 2006; Pircher et al. 2008; Naik et al. 2008). It was also anticipated that measuring EPCs in circulation would constitute an alternative approach to monitor the therapeutic efficacy of anti-angiogenesis treatment schedules (reviewed by DePrimo et al. 2007). Additionally, it has been demonstrated that in cardiovascular disease states, characterized by endothelial dysfunction (ED), there is a diminution in the availability and impairment of several EPCs functions (Vasa et al. 2001; Hill et al. 2003; Michowitz et al. 2007; Nonaka-Sarukawa et al. 2007). These progenitor cell associated-defects impair the efficient endogenous repair of the affected endothelium, being a potential indicator of ED and vascular disease severity. Concordantly, a direct quantification of circulating EPCs may represent a non-invasive surrogate marker for vascular homeostasis/dysfunction and cardiovascular disease (Michowitz et al. 2007; Nonaka-Sarukawa et al. 2007). In fact, reduced numbers of circulating EPCs were reported to predict future cardiovascular events (Schmidt-Lucke et al. 2005). As the Metabolic Syndrome (MS) comprises a group of risk factors for cardiovascular disease, a compelling link between MS and alterations in the biological features of EPCs has been demonstrated, and the use of EPCs as a novel MS disease biomarker is under evaluation (Fadini et al. 2006a).

6.2 MS-Associated Pathologies and EPCs Dysfunction – The Link Grows Stronger

6.2.1 *MS and EPCs*

MS is a highly prevalent condition in industrialized countries and represents a cluster of several risk factors for cardiovascular disease (see Chapter 1 for review). MS has been strongly associated with the development of insulin resistance and consequently of type 2 Diabetes Mellitus (T2DM) (reviewed by Soares and Costa 2007; Costa et al. 2007). To all these MS-related pathologies contribute the increase in oxidative stress, the presence of a systemic pro-inflammatory state and the occurrence of generalized ED (reviewed by Soares and Costa 2007; Costa et al. 2007; Suzuki et al. 2008). It has been previously established that endothelial damage ultimately represents a balance between the magnitude of injury and the capacity for endothelial repair (Hill et al. 2003). In the case of MS-related pathologies, the vasculogenic ability to restore the endothelium is compromised, originating a balance disruption and a shift towards vascular ED. This is thought to occur due to dysfunctions in EPCs biological activities, which by preventing the maintenance of endothelial monolayer integrity, sustain and exacerbate inflammatory and ED conditions. In metabolic disease it has been demonstrated that EPCs circulating levels are reduced (associated to impaired mobilization), their viability is affected, and their outgrowth functional capacity is defective, all these resulting in reduced regenerative potential (Fadini et al. 2006a, b). In addition, EPCs-associated alterations in MS are attributable to the accumulation of cardiovascular risk factors, associated with T2DM vascular alterations and atherosclerotic disease, validating the role of EPCs in the development of MS-related vascular complications (Tepper et al. 2002; Fadini et al. 2007; Satoh et al. 2008). Although the search for molecular connections between metabolic dysfunction and EPCs-functional impairment is ongoing, it is becoming apparent that the adipose tissue (AT) may play a relevant role in this process. Cumulative evidence has shown that the AT is far from being an inactive bystander and by releasing tissue-derived factors may modulate EPC recruitment and function (Wolk et al. 2005; Shibata et al. 2008). Nonetheless, increased glycaemic levels, insulin resistance, oxidative stress, dyslipidaemia and hypertension, feature conditions of MS, also crucially contribute to BM-derived EPCs malfunction, conducing to exacerbated vascular endothelial-associated dysfunction and vascular disease (Imanishi et al. 2004; Chen et al. 2007; Sorrentino et al. 2007; You et al. 2008b).

6.2.2 MS and AT-Derived Vasculogenesis Mediators

AT, is a type of loose connective tissue comprised of lipid-filled cells, the adipocytes, surrounded by a matrix of collagen fibers, blood vessels, fibroblasts and immune cells. As the result of its apparent structural and histological simplicity, AT functions were initially limited to energy storage, insulation, and thermoregulation (reviewed by Ahima and Flier 2000a). This concept has changed decades later after the discovery of the adipose-derived hormones adiponectin and leptin (Arita et al. 1999; Ahima and Flier 2000b). Nowadays, the AT is considered a dynamic endocrine organ, with the ability to synthesize and/or secrete a large number of enzymes, hormones, growth factors, components of the coagulation/fibrinolytic pathways, cytokines, complement factors, matrix and membrane proteins, collectively termed adipokines; which participate in a broad range of physiological processes, including systemic vasculogenesis (reviewed by Ahima and Flier 2000a; Wolk et al. 2005;

Frühbeck 2008). As mentioned in the previous chapter, adipocytes secrete a wide variety of pro-angiogenic molecules, which promote increased local angiogenic activity in AT, allowing the maintenance of MS-associated inflammatory conditions (Dallabrida et al. 2003; reviewed by Hausman and Richardson 2004; Silha et al. 2005). However, AT is suggested to regulate vasculogenesis in metabolic disease in a different fashion, by adipocyte-derived non-angiogenic factor production. It is thought that AT modulates EPCs biological features through the synthesis and release of the regulatory peptides, leptin and adiponectin (Ahima and Flier 2000b; Whitehead et al. 2006). In MS conditions, these AT-derived bioactive mediators were shown to change EPCs number and biological properties, altering vasculogenesis regenerative function and sustaining ED (Wolk et al. 2005; reviewed by Ribatti et al. 2007; Anagnostoulis et al. 2008; Mahadev et al. 2008).

6.2.3 Leptin and EPCs Function

Leptin is a pleiotropic hormone produced and released by adipocytes in response to paracrine, endocrine and neuroendocrine stimuli (Ahima and Flier 2000b). It is recognized that its secretion is proportional to the AT mass, being elevated in most obese subjects (You et al. 2008a). In addition to its primary role as controller of appetite and body weight, leptin, after binding to its specific receptors, has also other biological functions, including the regulation of postnatal vasculogenesis (Ahima and Flier 2000b; Wolk et al. 2005; Anagnostoulis et al. 2008). To date, a single report has established a connection between hyperleptinemia and EPC function, by showing that human EPCs express the leptin receptor and functionally respond to the hormone actions (Wolk et al. 2005). It was demonstrated that normal or elevated leptin levels, did not affect the differentiation or expansion potential of EPCs. At physiological concentrations, this hormone seemed to *in vitro* induce and increase the formation of EPCs-derived tubular networks. However, higher leptin levels impaired EPCs tubular-forming capacity, inhibiting also cell migration. Although further research is mandatory in order to better clarify the actions of leptin on EPCs recruitment and homing, this study has proposed a role for this hormone in the regulation of vasculogenesis, establishing also an association between hyperleptinemia, obesity and cardiovascular disease (Wolk et al. 2005). In fact, augmented levels of leptin and defective EPCs function have both been observed in MS patients with increased risk for cardiovascular events (Patel et al. 2008), suggesting an important connection between this AT-derived mediator and the regulation of vasculogenesis associated to vascular repair and cardiovascular function.

6.2.4 Adiponectin and EPCs Modulation

Adiponectin is an adipose-derived hormone, which plays a protective role in the development of obesity-linked diseases (Whitehead et al. 2006). This hormone seems to be the unique adipokine, whose secretion and circulating levels are inversely proportional to body fat content. Through engagement with its specific receptors, adiponectin exerts several beneficial effects, including the modulation adult vasculogenesis (Whitehead et al. 2006; reviewed by Ribatti et al. 2007; Shibata et al. 2008). In MS-related pathologies, such as coronary artery disease (CAD) and in T2DM, impairment in vascular remodelling has been associated to a downregulation of AT-derived adiponectin (Kumada et al. 2003). This observed vascular deficiency has been suggested to occur due to impaired EPCs mobilization and decreased circulation. In fact, low levels of adiponectin have been implicated in the alterations of EPCs biological features, contributing to vascular repair inhibition. It has been recently demonstrated that adiponectin has a positive regulation on human EPCs functions, by inducing their differentiation into matured ECs, having a stimulatory action on the development of EPCs-derived vascular networks and by promoting EPCs migration (Shibata et al. 2008). Even though it appears that this hormone can directly act on EPCs, mediating various crucial functions, many questions remain unanswered concerning the expression of adiponectin receptors by EPCs and their *in vivo* role in cell mobilization and recruitment in pathophysiological conditions.

Despite the modulator actions of AT-secreted adipokines in the regulation of postnatal vasculogenesis in MS, it is conceivable that the cumulative effects of the group of metabolic alterations including hyperglycaemia, insulin resistance, oxidative stress, dyslipidaemia and hypertension will, as a whole, impair the biological roles of EPCs.

6.3 MS-Associated Risk Factors – A Task Force Impairing Vasculogenesis in MS

The group of cardiovascular risk factors which comprises MS is thought to decisively induce vascular damages and to unable the efficient vasculogenic repair, predisposing the development of MS-related vasculopathy (Fadini et al. 2006a; Fadini et al. 2006b). The vasculogenic process in MS is extremely affected, as most MS-associated cardiovascular risk factors negatively influence the majority of EPCs biological functions, such as mobilization, viability, differentiation, clonogenic potential (number of colony forming units) and tubular formation capability; as will be further considered.

6.3.1 Hyperglycaemia

Vascular complications in T2DM are a significant cause of human morbidity and mortality, by affecting multiple organs, in particular the cardiovascular system, through the promotion of atherosclerosis (Nakagami et al. 2005). Hyperglycaemia is one of the major causal factors implicated in the development of vascular alterations (reviewed by Aronson 2008). High glucose levels are involved in the generation of advanced glycation end products (AGEs), which accumulate in the vessel wall, and that by interacting with its receptors (RAGE) induce oxidative stress, increased inflammation and ED (reviewed by Jandeleit-Dahm and Cooper 2008). Hyperglycaemia promotes the maintenance of ED conditions by directly impairing most EPCs-driven functional roles. Hyperglycaemia was been demonstrated to: decrease EPCs migration and integrative capacities (Kränkel et al. 2005); reduce the mobilization of EPCs (Gallagher et al. 2007); accelerate the onset of progenitor cell senescence (Chen et al. 2007); to inhibit EPCs colony-forming ability; to decrease the number and proliferation activity of both early and late EPCs (EOCs and LOCs) and to impair the migration and vasculogenesis activities of LOCs, the cell population suggested of having the "true" vasculogenic-associated morphogenesis capacity (Chen et al. 2007). Decreased vascular progenitor cells migration and inhibition of functional incorporation into tubular structures, were suggested to occur by hyperglycaemia-induced decrease in NO production and MMP-9 activity (Kränkel et al. 2005). Reduced mobilization of EPCs was showed to occur due to modifications in eNOS phosphorylation and activation status within the BM microenvironment, unabling efficient EPCs release from the marrow niche to the peripheral circulation (Gallagher et al. 2007; Ingram et al. 2008). The hyperglycaemia-induced EPCs senescence has been demonstrated to take place through multiple mechanisms such as, by promotion of telomere shortening, and through alterations in the p38 mitogen activated protein kinase (MAPK) and NO-mediated pathways (Kuki et al. 2006, Chen et al. 2007; Ingram et al. 2008). Recently, a novel molecular link has been discovered between high glucose levels and EPCs increased senescence, as it was demonstrated that the Sirtuin 1 (SIRT1) gene, which regulates cell cycle, premature senescence and apoptosis, is downregulated in EPCs. SIRT1 low expression levels impair the important cascade of intracellular events, culminating with EPCs early senescence (Balestrieri et al. 2008a). All these deleterious effects were reported to occur due to a direct effect of increased glucose levels on EPCs (Chen et al. 2007), however it has also been suggested that hyperglycaemia may promote EPCs dysfunction indirectly through the induction of overproduction of Reactive Oxygen Species (ROS) and increased oxidative stress (Callaghan et al. 2005). Importantly, anti-diabetic treatments were reported to improve re-endothelialization capacity of EPCs from diabetic individuals (Gensch et al. 2007).

6.3.2 Insulin Resistance

Approximately 80% of all T2DM coexist with insulin resistance (IR) (reviewed by Zimmet et al. 2001). Several studies have proposed that IR may affect unfavourably the balance between endothelial injury and endogenous repair, promoting ED and contributing to premature atherosclerosis (Dandona et al. 2003; Dandona et al. 2004). Apparently, IR aids ED perpectuation (Kim et al. 2006) by modulating vasculogenesis-associated EPCs capability of effectively promoting endothelium regeneration. Some of the EPCs biochemical alterations induced by IR are currently being evaluated. However, to date no direct links have been established between

IR and EPCs-induced dysfunctions. It is thought that EPCs biological modifications are mostly affected by IR in an indirect fashion, through the increase in ROS and by the activation of pro-inflammatory cytokines (Houstis et al. 2006; reviewed by Cubbon et al. 2007). In fact, IR states are closely linked to increased production of ROS, a characteristic feature of IR, and thought to play a causal role in its development (Houstis et al. 2006). The deleterious effects of oxidative stress in EPCs biological characteristics have been established and will be contemplated in section 6.3.5. Although a direct cause-effect has not been established between IR and EPCs alterations, it has been showed that treatments with insulin sensitizing drugs may improve EPCs functional parameters, independently of glycaemic levels and/or redox status (Schoonjans and Auwerx 2000). Nonetheless, further studies are necessary to clarify the molecular links between IR and vasculogenic impairment.

6.3.3 Dyslipidaemia

It has been suggested that dyslipidaemia increases the risk for atherosclerosis and CAD by inducing endothelial cell injury and dysfunction and inhibiting efficient regeneration by vasculogenesis (reviewed by Boak and Chin-Dusting 2004; Chen et al. 2004). It was reported that the number of vascular progenitor cells is significantly reduced in patients with hypercholesterolaemia and that functional activities of the isolated EPCs, such as proliferation, migration, adhesion and *in vitro* tubular forming capacity were impaired (Imanishi et al. 2003; Chen et al. 2004; Wang et al. 2004b; Imanishi et al. 2004). The aforementioned studies have only investigated the influence of low-density lipoprotein (LDL) cholesterol and none of them have addressed the role of high-density lipoprotein on EPCs number and function. It was then showed that oxidized-LDL (ox-LDL), one of the most important risk factors for cardiovascular disease (Holvoet et al. 2008), may play a relevant role in EPCs pathology. It has been shown *in vitro* that ox-LDL inhibited VEGF-induced EPCs differentiation through alterations on the phosphoinositide 3-kinase (PI3K)/Akt pathway (Imanishi et al. 2003). Additionally, ox-LDL was reported to increase EPCs senescence, through the inhibition of telomerase activity, consequently leading to an impairment of EPCs proliferative capacity and tubular network formation (Imanishi et al. 2004). Ox-LDL was also suggested to decrease eNOS protein expression in EPCs, promoting cell apoptosis and impairing its adhesive, migratory, and tube formation capability (Ma et al. 2006). In addition, ox-LDL may also induce EPCs apoptosis through the activation of the pro-apoptotic protein Bax (Jizhong et al. 2007). Besides the *in vitro* evidence, additional studies are required to clarify the *in vivo* effects of ox-LDL in EPCs functions. Nonetheless, as ox-LDL has several deleterious effects on the endothelial monolayer, including the impairment of eNO production and induction of superoxide anion formation, we may speculate that these alterations may modify the interaction of EPCs and the BM vasculature, preventing the efficient mobilization of EPCs from the BM into the circulation

(Kugiyama et al. 1990; Galle et al. 1995). Furthermore, it has been suggested that LDL lowering drugs, may ameliorate ED, by increasing NO bioavailability, reducing oxidant levels, inhibiting inflammatory responses, and concomitantly improving indirectly and/or directly EPC function (Li et al. 2008).

6.3.4 Hypertension

Elevated blood pressure levels are typically associated with significant mechanical endothelial injury and dysfunction (Spieker et al. 2000). Disruption of endothelial homeostasis in hypertensive patients is thought to worsen their cardiovascular prognosis (Perticone et al. 2001) and contribute to increase blood pressure levels (Schiffrin 2001). Alterations in postnatal vasculogenesis have been associated to hypertension-induced ED (You et al. 2008b; Watson et al. 2008). A clinical study in patients with CAD identified hypertension as a major independent risk factor predictor for impaired EPCs migration (Vasa et al. 2001). It was reported that the functional activity of EPCs is reduced in experimental model settings and in hypertensive patients, due to increased EPC-induced senescence (Imanishi et al. 2005). Although it is still unclear, hypertension does not seem to have a direct action on EPCs reduction of half-life, which may be caused by telomerase inactivation related to the increase in oxidative stress associated with hypertension (Higashi et al. 2002; Imanishi et al. 2005; Touyz et al. 2004). Further, studies have shown that in patients with arterial hypertension, no association was observed between the number of circulating vascular progenitor cells and hypertension, suggesting that cell mobilization may not be affected (Werner et al. 2005; Delva et al. 2007). Contradictorily, it was also reported that reduced levels of circulating $CD34^{+}VEGFR\text{-}2^{+}$ EPCs were detected in hypertensive patients as compared to normotensive individuals (Pirro et al. 2007). Lower levels of peripheral EPCs correlated with a downregulation in the homeobox A9 (HOXA9) gene expression, which is critical for endothelial commitment during progenitor cell maturation (Pirro et al. 2007). Although further studies are required in order to further clarify the role of hypertension in EPCs functions, it has been reported that antihypertensive drugs may improve vascular function through EPC activation (Yao et al. 2007).

6.3.5 Oxidative Stress

Increased oxidative stress has been proposed as an important molecular mechanism for vascular complications associated with DM, IR, hyperlipideamia and hypertension (reviewed in Chapter 3, Aronson 2008; Houstis et al. 2006; Francois and Kojda 2004; Yanai et al. 2008), by exerting a direct cytotoxic effect on the vascular monolayer (Griendling et al. 2003). ROS may directly harm the vascular endothelium while superoxide reacts with NO to form peroxynitrite anion

(ONOO−), a powerful oxidant (Griendling et al. 2003; Kuzkaya et al. 2003). Diminished release of eNO caused either by excessive oxidative degradation or impaired local production has been implicated in endothelial lining damage and insufficient repair capability due to deficient EPCs mobilization/functional status (Creager et al. 2003; Yao et al. 2006). In fact, oxidative stress-induced reduction of NO bioavailability represents the major mechanism leading to impaired EPCs *in vivo* re-endothelialization capacity and *in vitro* function (Sorrentino et al. 2007). NO deficient release by the vasculature is thought to alter EPCs migratory function and colony-forming ability, indicating a central role for eNO activity in EPC biology in oxidative stress conditions (Hill et al. 2003). Despite the crucial role played by NO, it has been recently shown in patients with MS and CAD that oxidative stress may directly induce DNA damage on EPCs, by promoting telomere shortening with consequent increase in vascular progenitor senescence rate, contributing to the progression of atherosclerosis (Satoh et al. 2008). Accordingly, it is plausible that the vasculature and the vasculogenic repair mechanism would benefit from antioxidant drugs, which exert cellular protective effects both by directly scavenging ROS reducing their damaging action and/or by potentially improving EPCs function.

6.4 Therapeutic Approaches in MS-Related Pathologies and Their Role on EPCs Function

ED is a recognized marker for atherosclerosis and cardiovascular disease; therefore, it is conceivable that therapies, which control ED-associated factors, namely hyperglycaemia, IR, dyslipidaemia, hypertension and oxidative stress, may improve endothelial function. In fact, vasculoprotective agents were suggested to recover endothelial monolayer functional properties by improving EPCs biological features, and preventing the progression of atherosclerosis. Several therapies were shown to promote the mobilization of endogenous EPCs; increasing their circulating number and recovering their biological functions, therefore ameliorating ED in cardiovascular disease progression by stimulation of vasculogenesis (reviewed by Werner and Nickenig 2006). It has been suggested that thiazolidinediones, peroxisome proliferators-activated receptor- gamma (PPARγ) agonists, by modulating the transcription of insulin-sensitive genes, lower serum glucose levels in patients with T2DM (Schoonjans and Auwerx 2000). In addition to their insulin-sensitizing effects, increasing evidence suggests that these drugs improve endothelium-dependent vascular function and prevent atherosclerotic disease progression, by restoring EPCs properties and actions. *In vitro* and *in vivo* studies have shown that PPARγ agonists may have direct effects on EPCs characteristics by increasing vascular progenitor cell proliferation, migratory activity and colony formation (Pistrosch et al. 2005; Wang et al. 2006; Werner et al. 2007); promoting EPCs differentiation and attenuating apoptosis (Wang et al. 2004a; Gensch et al. 2007). Further, 3-hydroxy-3-methylglutaryl coenzyme A (HMG-CoA) reductase inhibitors (statins) efficiently reduce plasma LDL cholesterol levels, and were shown to improve endothelial function also by modulating EPCs biochemical properties. These actions were though to be mediated, at least partly, by mechanisms independent of statins lipid lowering actions (pleiotropic effects) (reviewed by Kusuyama et al. 2006). *In vitro* studies suggested that statins improve EPCs proliferation, migration and survival functions by increasing the phosphorylation of eNOS via activation of the PI3K/Akt signalling pathway (Llevadot et al. 2001; Li et al. 2008). Experimental and clinical reports have shown that statins significantly induce the mobilization of EPCs, increasing the number of peripheral circulating cells and augmenting serum concentration of VEGF, stimulating also EPCs incorporation and re-endothelization properties (Llevadot et al. 2001; Walter et al. 2002; Li et al. 2008). Additionally, it has been observed that statins induce peripheral mobilization of both early and late EPCs, predominantly recruiting and maintaining in circulation late vascular progenitor cells (Deschaseaux et al. 2007). A role in ED reversion by vasculogenesis stimulation has also been attributed to angiotensin converting enzyme (ACE) inhibitors, and angiotensin receptor blockers (ARB), which thereby also reduce atherosclerosis and risk of cardiovascular events

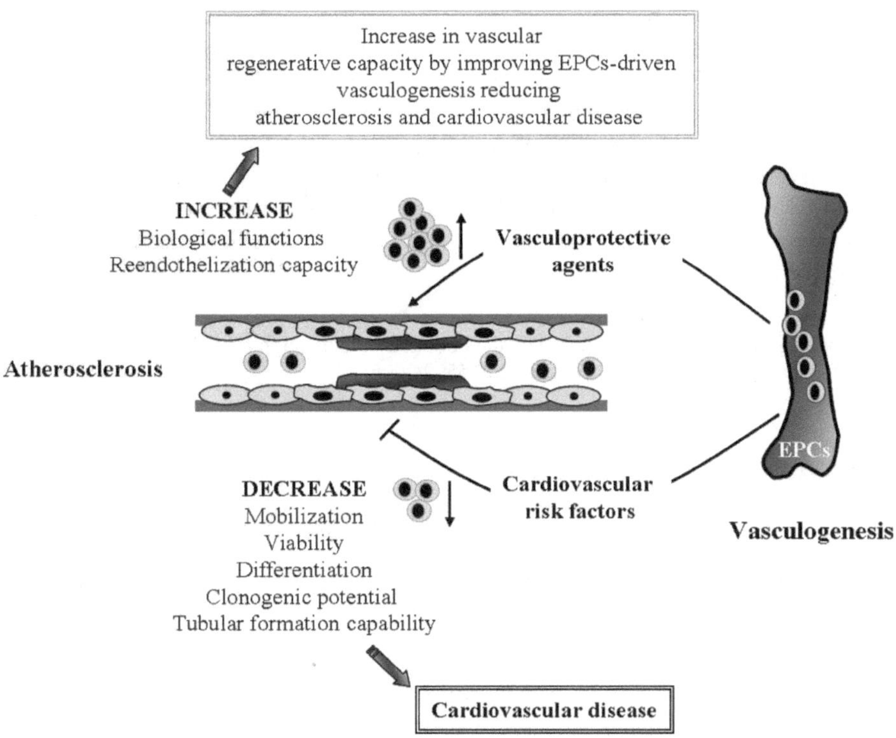

Fig. 6.1 Cardiovascular risks factors decrease EPCs number and function being associated with atherosclerotic disease. Vasculoprotective agents increase EPCs actions improving endothelial function and preventing the progression of atherosclerosis. Adapted from Costa and Vendeira, 2007

(Min et al. 2004). The aforementioned therapies besides decreasing systolic blood pressure in hypertensive settings have also shown beneficial effects on EPCs functions such as recruitment, proliferation, migration, and *in vitro* increased clonogenic potential (Min et al. 2004; Yao et al. 2007). EPCs modulator activities observed were mostly associated with therapy-induced antioxidant effects (Yao et al. 2007; reviewed by Werner and Nickenig 2006). In fact, the beneficial role of antioxidants in vascular biology and in the vasculogenesis mechanism is being currently addressed. It has been suggested that antioxidants such as vitamin C and E and different polyphenols may increase EPCs number, augment VEGF levels and induce p38 expression levels (Fiorito et al. 2008; Balestrieri et al. 2008b, c; also see Chapters 3 and 8).

These therapeutic approaches strongly suggest that strengthening the vascular regenerative capacity by improving EPCs-driven vasculogenesis may be an important path to reduce the incidence of atherosclerosis and cardiovascular disease associated to MS (Fig. 6.1).

6.5 Conclusions

Adult vasculogenesis has proven to be a crucial mechanism for improving ED and vascular endothelial recovery associated with cardiovascular risk factor injury. The majority of EPCs biological functions seem to be severely affected by the group of cardiovascular risk factors which comprises MS. The therapeutic control of these deleterious factors along with a better understanding of EPCs biochemical features may reveal essential in the prevention of the progression of atherosclerosis and cardiovascular disease. Therefore, the identification of novel compounds able to enhance endogenous EPC levels and to improve their functional activity will be of noticeable clinical interest.

Acknowledgments CC was supported by the Portuguese Foundation for Science and Technology (SFRH/BPD/40554/2007, PTDC/SAU-OSM/65599/2006)

References

Ahima RS, Flier JS. *Adipose tissue as an endocrine organ.* Trends Endocrinol Metab. 2000a; 11: 327–32.

Ahima RS, Flier JS. *Leptin.* Annu Rev Physiol. 2000b; 62: 413–37.

Aicher A, Heeschen C, Mildner-Rihm C, Urbich C, Ihling C, Technau-Ihling K, Zeiher AM, Dimmeler S. *Essential role of endothelial nitric oxide synthase for mobilization of stem and progenitor cells.* Nat Med. 2003; 9: 1370–6.

Aicher A, Zeiher AM, Dimmeler S. *Mobilizing endothelial progenitor cells.* Hypertension. 2005; 45: 321–5.

Anagnostoulis S, Karayiannakis AJ, Lambropoulou M, Efthimiadou A, Polychronidis A, Simopoulos C. *Human leptin induces angiogenesis in vivo.* Cytokine. 2008; 42: 353–7.

Arita Y, Kihara S, Ouchi N, Takahashi M, Maeda K, Miyagawa J, Hotta K, Shimomura I, Nakamura T, Miyaoka K, Kuriyama H, Nishida M, Yamashita S, Okubo K, Matsubara K,

Muraguchi M, Ohmoto Y, Funahashi T, Matsuzawa Y. *Paradoxical decrease of an adipose-specific protein, adiponectin, in obesity.* Biochem Biophys Res Commun. 1999; 257: 79–83.

Aronson D. *Hyperglycemia and the pathobiology of diabetic complications.* Adv Cardiol. 2008; 45: 1–16.

Asahara T, Murohara T, Sullivan A, Silver M, van der Zee R, Li T, Witzenbichler B, Schatteman G, Isner JM. *Isolation of putative progenitor endothelial cells for angiogenesis.* Science. 1997; 275: 964–7.

Asahara T, Takahashi T, Masuda H, Kalka C, Chen D, Iwaguro H, Inai Y, Silver M, Isner JM. *VEGF contributes to postnatal neovascularization by mobilizing bone marrow-derived endothelial progenitor cells.* EMBO J. 1999a; 18: 3964–72.

Asahara T, Masuda H, Takahashi T, Kalka C, Pastore C, Silver M, Kearne M, Magner M, Isner JM. *Bone marrow origin of endothelial progenitor cells responsible for postnatal vasculogenesis in physiological and pathological neovascularization.* Circ Res. 1999b; 85: 221–8.

Balestrieri ML, Rienzo M, Felice F, Rossiello R, Grimaldi V, Milone L, Casamassimi A, Servillo L, Farzati B, Giovane A, Napoli C. *High glucose downregulates endothelial progenitor cell number via SIRT1.* Biochim Biophys Acta. 2008a; 1784: 936–45.

Balestrieri ML, Fiorito C, Crimi E, Felice F, Schiano C, Milone L, Casamassimi A, Giovane A, Grimaldi V, del Giudice V, Minucci PB, Mancini FP, Servillo L, D'Armiento FP, Farzati B, Napoli C. *Effect of red wine antioxidants and minor polyphenolic constituents on endothelial progenitor cells after physical training in mice.* Int J Cardiol. 2008b; 126: 295–7.

Balestrieri ML, Schiano C, Felice F, Casamassimi A, Balestrieri A, Milone L, Servillo L, Napoli C. *Effect of low doses of red wine and pure resveratrol on circulating endothelial progenitor cells.* J Biochem. 2008c; 143: 179–86.

Boak L, Chin-Dusting JP. *Hypercholesterolemia and endothelium dysfunction: role of dietary supplementation as vascular protective agents.* Curr Vasc Pharmacol. 2004; 2: 45–52.

Callaghan MJ, Ceradini DJ, Gurtner GC. *Hyperglycemia-induced reactive oxygen species and impaired endothelial progenitor cell function.* Antioxid Redox Signal. 2005; 7: 1476–82.

Chen JZ, Zhang FR, Tao QM, Wang XX, Zhu JH, Zhu JH. *Number and activity of endothelial progenitor cells from peripheral blood in patients with hypercholesterolaemia.* Clin Sci. 2004; 107: 273–80.

Chen YH, Lin SJ, Lin FY, Wu TC, Tsao CR, Huang PH, Liu PL, Chen YL, Chen JW. *High glucose impairs early and late endothelial progenitor cells by modifying nitric oxide-related but not oxidative stress-mediated mechanisms.* Diabetes. 2007; 56: 1559–68.

Costa C, Incio J, Soares R. *Angiogenesis and chronic inflammation: cause or consequence?* Angiogenesis. 2007; 10: 149–66.

Costa C, Vendeira P. *Penis and endothelium – Extra genital aspects of erectile dysfunction.* Rev Int Androl. 2007; 5: 50–8.

Creager MA, Luscher TF, Cosentino F, Beckman JA. *Diabetes and vascular disease: pathophysiology, clinical consequences, and medical therapy, part I.* Circulation. 2003; 108: 1527–32.

Cubbon RM, Rajwani A, Wheatcroft SB. *The impact of insulin resistance on endothelial function, progenitor cells and repair.* Diab Vasc Dis Res. 2007; 4: 103–11.

Dallabrida SM, Zurakowski D, Shih SC, Smith LE, Folkman J, Moulton KS, Rupnick MA. *Adipose tissue growth and regression are regulated by angiopoietin-1.* Biochem Biophys Res Commun. 2003; 311: 563–71.

Dandona P, Aljada A, Chaudhuri A, Bandyopadhyay A. *The potential influence of inflammation and insulin resistance on the pathogenesis and treatment of atherosclerosis-related complications in type 2 diabetes.* J Clin Endocrinol Metab. 2003; 88: 2422–9.

Dandona P, Aljada A, Bandyopadhyay A. *Inflammation: the link between insulin resistance, obesity and diabetes.* Trends Immunol. 2004; 25: 4–7.

De Palma M, Venneri MA, Galli R, Sergi L, Politi LS, Sampaolesi M, Naldini L. *Tie2 identifies a hematopoietic lineage of proangiogenic monocytes required for tumor vessel formation and a mesenchymal population of pericyte progenitors.* Cancer Cell. 2005; 8: 211–26.

Delva P, Degan M, Vallerio P, Arosio E, Minuz P, Amen G, Di Chio M, Lechi A. *Endothelial progenitor cells in patients with essential hypertension.* J Hypertens. 2007; 25: 127–32.

DePrimo SE, Bello C. *Surrogate biomarkers in evaluating response to anti-angiogenic agents: focus on sunitinib.* Ann Oncol. 2007; 18 Suppl 10: x11–9.

Deschaseaux F, Selmani Z, Falcoz PE, Mersin N, Meneveau N, Penfornis A, Kleinclauss C, Chocron S, Etievent JP, Tiberghien P, Kantelip JP, Davani S. *Two types of circulating endothelial progenitor cells in patients receiving long term therapy by HMG-CoA reductase inhibitors.* Eur J Pharmacol. 2007; 562: 111–8.

Fadini GP, de Kreutzenberg SV, Coracina A, Baesso I, Agostini C, Tiengo A, Avogaro A. *Circulating CD34+ cells, metabolic syndrome, and cardiovascular risk.* Eur Heart J. 2006a; 27: 2247–55.

Fadini GP, Sartore S, Albiero M, Baesso I, Murphy E, Menegolo M, Grego F, Vigili de Kreutzenberg S, Tiengo A, Agostini C, Avogaro A. *Number and function of endothelial progenitor cells as a marker of severity for diabetic vasculopathy.* Arterioscler Thromb Vasc Biol. 2006b; 26: 2140–6.

Fadini GP, Agostini C, Sartore S, Avogaro A. *Endothelial progenitor cells in the natural history of atherosclerosis.* Atherosclerosis. 2007; 194:46–54.

Fiorito C, Rienzo M, Crimi E, Rossiello R, Luisa Balestrieri M, Casamassimi A, Muto F, Grimaldi V, Giovane A, Farzati B, Mancini FP, Napoli C. *Antioxidants increase number of progenitor endothelial cells through multiple gene expression pathways.* Free Radic Res. 2008; 42: 754–62.

Folkman J. *What is the role of endothelial cells in angiogenesis?* Lab Invest. 1984; 51: 601–4.

Francois M, Kojda G. *Effect of hypercholesterolemia and of oxidative stress on the nitric oxide-cGMP pathway.* Neurochem Int. 2004; 45: 955–61.

Frühbeck G. *Overview of adipose tissue and its role in obesity and metabolic disorders.* Methods Mol Biol. 2008; 456: 1–22.

Gallagher KA, Liu ZJ, Xiao M, Chen H, Goldstein LJ, Buerk DG, Nedeau A, Thom SR, Velazquez OC. *Diabetic impairments in NO-mediated endothelial progenitor cell mobilization and homing are reversed by hyperoxia and SDF-1 alpha.* J Clin Invest. 2007; 117: 1249–59.

Galle J, Bengen J, Schollmeyer P, Wanner C. *Impairment of endothelium-dependent dilation in rabbit renal arteries by oxidized lipoprotein (a): Role of oxygen-derived radicals.* Circulation. 1995; 92: 1582–9.

Gensch C, Clever YP, Werner C, Hanhoun M, Böhm M, Laufs U. *The PPAR-gamma agonist pioglitazone increases neoangiogenesis and prevents apoptosis of endothelial progenitor cells.* Atherosclerosis. 2007; 192: 67–74.

Gill M, Dias S, Hattori K, Rivera ML, Hicklin D, Witte L, Girardi L, Yurt R, Himel H, Rafii S. *Vascular trauma induces rapid but transient mobilization of VEGFR2(+)AC133(+) endothelial precursor cells.* Circ Res. 2001; 88: 167–74.

Griendling KK, FitzGerald GA. *Oxidative stress and cardiovascular injury. Part II. Animal and human studies.* Circulation. 2003; 108: 2034–40.

Grunewald M, Avraham I, Dor Y, Bachar-Lustig E, Itin A, Jung S, Chimenti S, Landsman L, Abramovitch R, Keshet E. *VEGF-induced adult neovascularization: recruitment, retention, and role of accessory cells.* Cell. 2006; 124:175–89.

Gulati R, Jevremovic D, Peterson TE, Chatterjee S, Shah V, Vile RG, Simari RD. *Diverse origin and function of cells with endothelial phenotype obtained from adult human blood.* Circ Res. 2003; 93: 1023–5.

Hausman GJ, Richardson RL. *Adipose tissue angiogenesis.* J Anim Sci. 2004; 82: 925–34.

Heissig B, Hattori K, Dias S, Friedrich M, Ferris B, Hackett NR, Crystal RG, Besmer P, Lyden D, Moore MA, Werb Z, Rafii S. *Recruitment of stem and progenitor cells from the bone marrow niche requires MMP-9 mediated release of kit-ligand.* Cell. 2002; 109: 625–37.

Higashi Y, Sasaki S, Nakagawa K, et al. *Endothelial function and oxidative stress in renovascular hypertension.* N Engl J Med. 2002; 346: 1954–62.

Hill JM, Zalos G, Halcox JP, Schenke WH, Waclawiw MA, Quyyumi AA, Finkel T. *Circulating endothelial progenitor cells, vascular function, and cardiovascular risk.* N Engl J Med. 2003; 348: 593–00.

Ho JW, Pang RW, Lau C, Sun CK, Yu WC, Fan ST, Poon RT. *Significance of circulating endothelial progenitor cells in hepatocellular carcinoma.* Hepatology. 2006; 44: 836–43.

Holvoet P, Lee DH, Steffes M, Gross M, Jacobs DR Jr. *Association between circulating oxidized low-density lipoprotein and incidence of the metabolic syndrome.* JAMA. 2008; 299: 2287–93.

Houstis N, Rosen ED, Lander ES. *Reactive oxygen species have a causal role in multiple forms of insulin resistance.* Nature. 2006; 440: 944–8.

Hristov M, Erl W, Weber PC. *Endothelial progenitor cells: isolation and characterization.* Trends Cardiovasc Med. 2003; 13: 201–6.

Hur J, Yoon CH, Kim HS, Choi JH, Kang HJ, Hwang KK, Oh BH, Lee MM, Park YB. *Characterization of two types of endothelial progenitor cells and their different contributions to neovasculogenesis.* Arterioscler Thromb Vasc Biol. 2004; 24: 288–93.

Imanishi T, Hano T, Matsuo Y, Nishio I. *Oxidized low-density lipoprotein inhibits vascular endothelial growth factor-induced endothelial progenitor cell differentiation.* Clin Exp Pharmacol Physiol. 2003; 30: 665–70.

Imanishi T, Hano T, Sawamura T, Nishio I. *Oxidized low-density lipoprotein induces endothelial progenitor cell senescence, leading to cellular dysfunction.* Clin Exp Pharmacol Physiol. 2004; 31: 407–13.

Imanishi T, Moriwaki C, Hano T, Nishio I. *Endothelial progenitor cell senescence is accelerated in both experimental hypertensive rats and patients with essential hypertension.* J Hypertens. 2005; 23: 1831–7.

Ingram DA, Lien IZ, Mead LE, Estes M, Prater DN, Derr-Yellin E, DiMeglio LA, Haneline LS. *In vitro hyperglycemia or a diabetic intrauterine environment reduces neonatal endothelial colony-forming cell numbers and function.* Diabetes. 2008; 57: 724–31.

Ishikawa M, Asahara T. *Endothelial progenitor cell culture for vascular regeneration.* Stem Cells Dev. 2004; 13: 344–9.

Jandeleit-Dahm K, Cooper ME. *The role of AGEs in cardiovascular disease.* Curr Pharm Des. 2008; 14: 979–86.

Jizhong C, Ruwen C, Chu-Huang C, Jie D. *Oxidized low-density lipoprotein stimulates p53-dependent activation of proapoptotic bax leading to apoptosis of differentiated endothelial progenitor cells.* Endocrinology. 2007; 148: 2085–94.

Kaplan RN, Riba RD, Zacharoulis S, Bramley AH, Vincent L, Costa C, MacDonald DD, Jin DK, Shido K, Kerns SA, Zhu Z, Hicklin D, Wu Y, Port JL, Altorki N, Port ER, Ruggero D, Shmelkov SV, Jensen KK, Rafii S, Lyden D. *VEGFR1-positive haematopoietic bone marrow progenitors initiate the pre-metastatic niche.* Nature. 2005; 438: 820–7.

Kim JA, Montagnani M, Koh KK, Quon MJ. *Reciprocal relationships between insulin resistance and endothelial dysfunction: molecular and pathophysiological mechanisms.* Circulation. 2006; 113: 1888–04.

Kränkel N, Adams V, Linke A, Gielen S, Erbs S, Lenk K, Schuler G, Hambrecht R. *Hyperglycemia reduces survival and impairs function of circulating blood-derived progenitor cells.* Arterioscler Thromb Vasc Biol. 2005; 25: 698–03.

Kugiyama K, Kerns SA, Morrisett JD, Roberts R, Henry PD. *Impairment of endothelium-dependent arterial relaxation by lysolecithin in modified low-density lipoproteins.* Nature. 1990; 344: 160–2.

Kuki S, Imanishi T, Kobayashi K, Matsuo Y, Obana M, Akasaka T. *Hyperglycemia accelerated endothelial progenitor cell senescence via the activation of p38 mitogen-activated protein kinase.* Circ J. 2006; 70: 1076–81.

Kumada M, Kihara S, Sumitsuji S, Kawamoto T, Matsumoto S, Ouchi N, Arita Y, Okamoto Y, Shimomura I, Hiraoka H, Nakamura T, Funahashi T, Matsuzawa Y; Osaka CAD Study Group. *Coronary artery disease. Association of hypoadiponectinemia with coronary artery disease in men.* Arterioscler Thromb Vasc Biol. 2003; 23: 85–9.

Kusuyama T, Omura T, Nishiya D, Enomoto S, Matsumoto R, Murata T, Takeuchi K, Yoshikawa J, Yoshiyama M. *The effects of HMG-CoA reductase inhibitor on vascular progenitor cells.* J Pharmacol Sci. 2006; 101: 344–9.

Kuzkaya N, Weissmann N, Harrison DG, Dikalov S. *Interactions of peroxynitrite, tetrahydrobiopterin, ascorbic acid, and thiols: implications for uncoupling endothelial nitric-oxide synthase.* J Biol Chem. 2003; 278: 22546–54.

Li X, Xu B. *HMG-CoA reductase inhibitor regulates endothelial progenitor function through the phosphatidylinositol 3'-Kinase/AKT signal transduction pathway.* Appl Biochem Biotechnol. 2008 Jun 18. [Epub ahead of print]

Llevadot J, Murasawa S, Kureishi Y, Uchida S, Masuda H, Kawamoto A, et al. *HMG-CoA reductase inhibitor mobilizes bone marrow – derived endothelial progenitor cells.* J Clin Invest. 2001; 108: 399–05.

Lyden D, Hattori K, Dias S, Costa C, Blaikie P, Butros L, Chadburn A, Heissig B, Marks W, Witte L, Wu Y, Hicklin D, Zhu Z, Hackett NR, Crystal RG, Moore MA, Hajjar KA, Manova K, Benezra R, Rafii S. *Impaired recruitment of bone-marrow-derived endothelial and hematopoietic precursor cells blocks tumor angiogenesis and growth.* Nat Med. 2001; 7: 1194–01.

Ma FX, Zhou B, Chen Z, Ren Q, Lu SH, Sawamura T, Han ZC. *Oxidized low density lipoprotein impairs endothelial progenitor cells by regulation of endothelial nitric oxide synthase.* J Lipid Res. 2006; 47: 1227–37.

Mahadev K, Wu X, Donnelly S, Ouedraogo R, Eckhart AD, Goldstein BJ. *Adiponectin inhibits vascular endothelial growth factor-induced migration of human coronary artery endothelial cells.* Cardiovasc Res. 2008; 78: 376–84.

Matsumoto T, Mifune Y, Kawamoto A, Kuroda R, Shoji T, Iwasaki H, Suzuki T, Oyamada A, Horii M, Yokoyama A, Nishimura H, Lee SY, Miwa M, Doita M, Kurosaka M, Asahara T. *Fracture induced mobilization and incorporation of bone marrow-derived endothelial progenitor cells for bone healing.* J Cell Physiol. 2008; 215: 234–42.

Michowitz Y, Goldstein E, Wexler D, Sheps D, Keren G, George J. *Circulating endothelial progenitor cells and clinical outcome in patients with congestive heart failure.* Heart. 2007; 93: 1046–50.

Min TQ, Zhu CJ, Xiang WX, Hui ZJ, Peng SY. *Improvement in endothelial progenitor cells from peripheral blood by ramipril therapy in patients with stable coronary artery disease.* Cardiovasc Drugs Ther. 2004; 18: 203–9.

Naik RP, Jin D, Chuang E, Gold EG, Tousimis EA, Moore AL, Christos PJ, de Dalmas T, Donovan D, Rafii S, Vahdat LT. *Circulating endothelial progenitor cells correlate to stage in patients with invasive breast cancer.* Breast Cancer Res Treat. 2008; 107: 133–8.

Nakagami H, Kaneda Y, Ogihara T, Morishita R. *Endothelial dysfunction in hyperglycemia as a trigger of atherosclerosis.* Curr Diabetes Rev. 2005; 1: 59–63.

Nonaka-Sarukawa M, Yamamoto K, Aoki H, Nishimura Y, Tomizawa H, Ichida M, Eizawa T, Muroi K, Ikeda U, Shimada K. *Circulating endothelial progenitor cells in congestive heart failure.* Int J Cardiol. 2007; 119: 344–8.

Patel SB, Reams GP, Spear RM, Freeman RH, Villarreal D. *Leptin: linking obesity, the metabolic syndrome, and cardiovascular disease.* Curr Hypertens Rep. 2008; 10: 131–7.

Peichev M, Naiyer AJ, Pereira D, et al. *Expression of VEGFR-2 and AC133 by circulating human CD34(+) cells identifies a population of functional endothelial precursors.* Blood. 2000; 95: 952–8.

Perticone F, Ceravolo R, Pujia A, Ventura G, Iacopino S, Scozzafava A, et al. *Prognostic significance of endothelial dysfunction in hypertensive patients.* Circulation. 2001; 104: 191–6.

Peters BA, Diaz LA, Polyak K, Meszler L, Romans K, Guinan EC, Antin JH, Myerson D, Hamilton SR, Vogelstein B, Kinzler KW, Lengauer C. *Contribution of bone marrow-derived endothelial cells to human tumor vasculature.* Nat Med. 2005; 11: 261–2.

Pircher A, Kähler CM, Skvortsov S, Dlaska M, Kawaguchi G, Schmid T, Gunsilius E, Hilbe W. *Increased numbers of endothelial progenitor cells in peripheral blood and tumor specimens in non-small cell lung cancer: a methodological challenge and an ongoing debate on the clinical relevance.* Oncol Rep. 2008; 19: 345–52.

Pirro M, Schillaci G, Menecali C, Bagaglia F, Paltriccia R, Vaudo G, Mannarino MR, Mannarino E. *Reduced number of circulating endothelial progenitors and HOXA9 expression in CD34+ cells of hypertensive patients.* J Hypertens. 2007; 25: 2093–9.

Pistrosch F, Herbrig K, Oelschlaegel U, Richter S, Passauer J, Fischer S, Gross P. *PPARgamma-agonist rosiglitazone increases number and migratory activity of cultured endothelial progenitor cells.* Atherosclerosis. 2005; 183: 163–7.

Rafii S. *Circulating endothelial precursors: mystery, reality, and promise.* J Clin Invest. 2000; 105: 17–9.

Ribatti D, Conconi MT, Nussdorfer GG. *Nonclassic endogenous novel regulators of angiogenesis.* Pharmacol Rev. 2007; 59: 185–05.

Risau W, Flamme I. *Vasculogenesis.* Annu Rev Cell Dev Biol. 1995; 11: 73–91.

Risau W. *Mechanisms of angiogenesis.* Nature. 1997; 386: 671–4.

Satoh M, Ishikawa Y, Takahashi Y, Itoh T, Minami Y, Nakamura M. *Association between oxidative DNA damage and telomere shortening in circulating endothelial progenitor cells obtained from metabolic syndrome patients with coronary artery disease.* Atherosclerosis. 2008; 198: 347–53.

Schiffrin EL. *A critical review of the role of endothelial factors in the pathogenesis of hypertension.* J Cardiovasc Pharmacol. 2001; 38: S3–6.

Schmidt-Lucke C, Rossig L, Fichtlscherer S, Vasa M, Britten M, Kamper U, Dimmeler S, Zeiher AM. *Reduced number of circulating endothelial progenitor cells predicts future cardiovascular events: proof of concept for the clinical importance of endogenous vascular repair.* Circulation. 2005; 111: 2981–7.

Schoonjans K, Auwerx J. *Thiazolidinediones: an update.* Lancet. 2000; 355: 1008–10.

Sepúlveda P, Martinez-León J, García-Verdugo JM. *Neoangiogenesis with endothelial precursors for the treatment of ischemia.* Transplant Proc. 2007; 39: 2089–94.

Shantsila E, Watson T, Lip GY. *Endothelial progenitor cells in cardiovascular disorders.* J Am Coll Cardiol. 2007; 49: 741–52.

Shi Q, Rafii S, Wu MH, Wijelath ES, Yu C, Ishida A, Fujita Y, Kothari S, Mohle R, Sauvage LR, Moore MA, Storb RF, Hammond WP. *Evidence for circulating bone marrow-derived endothelial cells.* Blood. 1998; 92: 362–7.

Shibata R, Skurk C, Ouchi N, Galasso G, Kondo K, Ohashi T, Shimano M, Kihara S, Murohara T, Walsh K. *Adiponectin promotes endothelial progenitor cell number and function.* FEBS Lett. 2008; 582: 1607–12.

Shintani S, Murohara T, Ikeda H, Ueno T, Honma T, Katoh A, Sasaki K, Shimada T, Oike Y, Imaizumi T. *Mobilization of endothelial progenitor cells in patients with acute myocardial infarction.* Circulation. 2001; 103: 2776–9.

Shmelkov SV, Butler JM, Hooper AT, Hormigo A, Kushner J, Milde T, St Clair R, Baljevic M, White I, Jin DK, Chadburn A, Murphy AJ, Valenzuela DM, Gale NW, Thurston G, Yancopoulos GD, D'Angelica M, Kemeny N, Lyden D, Rafii S. *CD133 expression is not restricted to stem cells, and both CD133+ and CD133- metastatic colon cancer cells initiate tumors.* J Clin Invest. 2008; 118: 2111–20.

Silha JV, Krsek M, Sucharda P, Murphy LJ. *Angiogenic factors are elevated in overweight and obese individuals.* Int J Obes. 2005; 29: 1308–14.

Smythe J, Fox A, Fisher N, Frith E, Harris AL, Watt SM. *Measuring angiogenic cytokines, circulating endothelial cells, and endothelial progenitor cells in peripheral blood and cord blood: VEGF and CXCL12 correlate with the number of circulating endothelial progenitor cells in peripheral blood.* Tissue Eng Part C Methods. 2008; 14: 59–67.

Soares R, Costa C. *Angiogenesis and inflammatory diseases: current concepts and therapeutic perspectives.* In: Maragoudakis ME; Papadimitriou E (ed.) Angiogenesis. Basic science and clinical applications, 1st edn. Transworld Research Network. 2007; 511–47.

Sorrentino SA, Bahlmann FH, Besler C, Müller M, Schulz S, Kirchhoff N, Doerries C, Horváth T, Limbourg A, Limbourg F, Fliser D, Haller H, Drexler H, Landmesser U. *Oxidant stress impairs in vivo reendothelialization capacity of endothelial progenitor cells from patients with type 2 diabetes mellitus: restoration by the peroxisome proliferator-activated receptor-gamma agonist rosiglitazone.* Circulation. 2007; 116: 163–73.

Spieker LE, Noll G, Ruschitzka FT, Maier W, Luscher TF. *Working under pressure: the vascular endothelium in arterial hypertension.* J Hum Hypertens. 2000; 14: 617–30.

Suzuki T, Hirata K, Elkind MS, Jin Z, Rundek T, Miyake Y, Boden-Albala B, Di Tullio MR, Sacco R, Homma S. *Metabolic syndrome, endothelial dysfunction, and risk of cardiovascular events: the Northern Manhattan Study (NOMAS).* Am Heart J. 2008; 156: 405–10.

Takahashi T, Kalka C, Masuda H, Chen D, Silver M, Kearney M, Magner M, Isner JM, Asahara T. *Ischemia- and cytokine-induced mobilization of bone marrow-derived endothelial progenitor cells for neovascularization.* Nat Med. 1999; 5: 434–8.

Tepper OM, Galiano RD, Capla JM, Kalka C, Gagne PJ, Jacobowitz GR, Levine JP, Gurtner GC. *Human endothelial progenitor cells from type II diabetics exhibit impaired proliferation, adhesion, and incorporation into vascular structures.* Circulation. 2002; 106: 2781–6.

Touyz RM, Schiffrin EL. *Reactive oxygen species in vascular biology: implications in hypertension.* Histochem Cell Biol. 2004; 122: 339–52.

Vasa M, Fichtlscherer S, Aicher A, Adler K, Urbich C, Martin H, Zeiher AM, Dimmeler S. *Number and migratory activity of circulating endothelial progenitor cells inversely correlate with risk factors for coronary artery disease.* Circ Res. 2001; 89: E1–7.

Walter DH, Rittig K, Bahlmann FH, Kirchmair R, Silver M, Murayama T, et al. *Statin therapy accelerates reendothelialization: a novel effect involving mobilization and incorporation of bone marrow-derived endothelial progenitor cells.* Circulation. 2002; 105: 3017–24.

Wang CH, Ciliberti N, Li SH, Szmitko PE, Weisel RD, Fedak PW, Al-Omran M, Cherng WJ, Li RK, Stanford WL, Verma S. *Rosiglitazone facilitates angiogenic progenitor cell differentiation toward endothelial lineage: a new paradigm in glitazone pleiotropy.* Circulation. 2004a; 109: 1392–00.

Wang CH, Ting MK, Verma S, Kuo LT, Yang NI, Hsieh IC, Wang SY, Hung A, Cherng WJ. *Pioglitazone increases the numbers and improves the functional capacity of endothelial progenitor cells in patients with diabetes mellitus.* Am Heart J. 2006; 152: 1051.e1–8.

Wang X, Chen J, Tao Q, Zhu J, Shang Y. *Effects of ox-LDL on number and activity of circulating endothelial progenitor cells.* Drug Chem Toxicol. 2004b; 27: 243–55.

Watson T, Goon PK, Lip GY. *Endothelial progenitor cells, endothelial dysfunction, inflammation, and oxidative stress in hypertension.* Antioxid Redox Signal. 2008; 10: 1079–88.

Werner C, Kamani CH, Gensch C, Böhm M, Laufs U. *The peroxisome proliferator-activated receptor-gamma agonist pioglitazone increases number and function of endothelial progenitor cells in patients with coronary artery disease and normal glucose tolerance.* Diabetes. 2007; 56: 2609–15.

Werner N, Kosiol S, Schiegl T, et al. *Circulating endothelial progenitor cells and cardiovascular outcomes.* N Engl J Med. 2005; 353: 999–07.

Werner N, Nickenig G. *Influence of cardiovascular risk factors on endothelial progenitor cells: limitations for therapy?* Arterioscler Thromb Vasc Biol. 2006; 26: 257–66.

Whitehead JP, Richards AA, Hickman IJ, Macdonald GA, Prins JB. *Adiponectin- a key adipokine in the metabolic syndrome.* Diabetes Obes Metab. 2006; 8: 264–80.

Wolk R, Deb A, Caplice NM, Somers VK. *Leptin receptor and functional effects of leptin in human endothelial progenitor cells.* Atherosclerosis. 2005; 183: 131–9.

Yanai H, Tomono Y, Ito K, Furutani N, Yoshida H, Tada N. *The underlying mechanisms for development of hypertension in the metabolic syndrome.* Nutr J. 2008; 7: 10.

Yao EH, Yu Y, Fukuda N. *Oxidative stress on progenitor and stem cells in cardiovascular diseases.* Curr Pharm Biotechnol. 2006; 7: 101–8.

Yao EH, Fukuda N, Matsumoto T, Kobayashi N, Katakawa M, Yamamoto C, Tsunemi A, Suzuki R, Ueno T, Matsumoto K. *Losartan improves the impaired function of endothelial progenitor cells in hypertension via an antioxidant effect.* Hypertens Res. 2007; 30: 1119–28.

Yoder MC, Mead LE, Prater D, Krier TR, Mroueh KN, Li F, Krasich R, Temm CJ, Prchal JT, Ingram DA. *Redefining endothelial progenitor cells via clonal analysis and hematopoietic stem/progenitor cell principals*. Blood. 2007; 109: 1801–9.

Yoon CH, Hur J, Park KW, Kim JH, Lee CS, Oh IY, Kim TY, Cho HJ, Kang HJ, Chae IH, Yang HK, Oh BH, Park YB, Kim HS. *Synergistic neovascularization by mixed transplantation of early endothelial progenitor cells and late outgrowth endothelial cells: the role of angiogenic cytokines and matrix metalloproteinases*. Circulation. 2005; 112: 1618–27.

You T, Nicklas BJ, Ding J, Penninx BW, Goodpaster BH, Bauer DC, Tylavsky FA, Harris TB, Kritchevsky SB. *The metabolic syndrome is associated with circulating adipokines in older adults across a wide range of adiposity*. J Gerontol A Biol Sci Med Sci. 2008a; 63: 414–9.

You D, Cochain C, Loinard C, Vilar J, Mees B, Duriez M, Lévy BI, Silvestre JS. *Hypertension impairs postnatal vasculogenesis: role of antihypertensive agents*. Hypertension. 2008b; 51: 1537–44.

Zimmet P, Alberti KG, Shaw J. *Global and societal implications of the diabetes epidemic*. Nature. 2001; 414: b782–7.

Chapter 7
Vascular Glucose Transport and the Metabolic Syndrome

Fatima Martel and Elisa Keating

Contents

Abstract The metabolic syndrome comprises a wide range of physiopathological abnormalities, all involved in systemic changes, occurring in different territories of the body. Glucose transporter changes throughout the body are observed in diverse degrees in the metabolic syndrome. This chapter aims to describe the alterations of glucose transport associated with the metabolic syndrome occurring at the vascular

---

F. Martel (✉)
Department of Biochemistry (U38-FCT), Faculty of Medicine, University of Porto, Al. Prof. Hernâni Monteiro, 4200-319 Porto, Portugal
e-mail: fmartel@med.up.pt

R. Soares, C. Costa (eds.), *Oxidative Stress, Inflammation and Angiogenesis in the Metabolic Syndrome*, DOI 10.1007/978-1-4020-9701-0_7,

level. We will highlight the mechanisms regulating glucose transport in endothelial and vascular smooth muscle cells derived from peripheral vascular beds and from the blood-brain and blood-retinal barriers. Knowledge on this subject will contribute for understanding the physiopathology of the metabolic syndrome, as well as possible treatments for this condition.

Keywords Diabetes · Endothelium · Glucose transport · Metabolic syndrome · Vascular smooth muscle

7.1 Introduction

The endothelial cell monolayer in blood vessels forms a barrier that contains the blood within vessel lumen, and provides a barrier for the exchange of nutrients and other substances. It is itself also actively involved in the local control of vascular homeostasis (eg. it forms an antithrombotic surface due to surface expression of molecules, such as heparan sulphate, and through the synthesis and release of anti-thrombogenic substances such as prostacyclins). When the vascular endothelial cell (VEC) monolayer is injured, its anti-thrombotic and dilatory properties are compromised, along with its effectiveness to form an impermeable barrier to blood cells. Moreover, injured endothelium secretes increased amounts of chemotactic factors that attract monocytes and smooth muscle cells to the intima layer to form early atherosclerotic plaques.

This chapter aims to review the effect of some metabolic syndrome-related abnormalities (hyperglycaemia, diabetes, insulin, hypoxia, nitric oxide (NO), etc) on glucose transporters in peripheral vascular beds and the blood-brain and blood-retinal barriers. We have focused principally on transport processes in endothelial cells but have also reviewed the available literature for vascular smooth muscle cells.

7.2 Metabolic Syndrome and Vascular Dysfunction

The metabolic syndrome (MS) is a constellation of metabolic risk factors and physical conditions that are accompanied by an enhanced propensity toward the development of type-2 diabetes mellitus (DM), atherosclerosis, and cardiovascular disease. The two major underlying risk factors for the MS are obesity and insulin resistance (reviewed in Chapter 1). Exacerbating factors are physical inactivity, advancing age, and endocrine and genetic factors. Associated hyperinsulinaemia, hyperglycaemia, and elevated adipokine levels lead to vascular endothelial dysfunction, an abnormal lipid profile, hypertension, and vascular inflammation, all of which promote the development of atherosclerotic cardiovascular disease.

Endothelial dysfunction (ED) comprises an array of defects in the behaviour of the endothelium, being usually evaluated by the reactivity of the endothelium to vasodilating or vasoconstricting stimuli. ED starts long before the development of structural atherosclerosis, and has gained considerable interest over the last years, especially as a predictive factor for future cardiovascular events. Recent insights into vascular biology enable us to understand the molecular mechanisms underlying ED. Diminished production of NO and/or increased inactivation of NO through oxidative stress [reactive oxygen species (ROS) and reactive nitrogen species (RNS)] are the basis of ED (Wiernsperger et al. 2007).

7.3 Diabetes and Vascular Dysfunction

Micro- and macrovascular diseases are serious complications of type-1 (insulin-dependent) and type-2 (non-insulin dependent) diabetes, constituting the main causes of morbidity and mortality in the diabetic population (Grundy et al. 1999). Microvascular dysfunction often results in severe pathologies, such as retinopathy or nephropathy, whereas macrovascular dysfunction leads to an accelerated development of endothelial cell dysfunction and atherosclerosis (Schalkwijk and Stehouwer 2005).

Chronic hyperglycaemia is considered a major risk factor for the development of both types of vascular diseases (Stratton et al. 2000; Wei et al. 1998; Yamagishi et al. 2007; Yu and Lyons 2005). Recent studies suggest that post-prandial hyperglycaemic excursions also adversely affect the vascular wall (Yamagishi et al. 2007). Other independent risk factors such as hyperlipidaemia, obesity or hypertension, also participate in the aetiology of cardiovascular disease (Yu and Lyons 2005).

Several facts explain the hyperglycaemia-induced vascular dysfunction. Among them, it has been suggested that the combination of hyperglycaemia-induced impeded proliferation and increased apoptosis of VEC is one of the earliest events in lesion formation in blood vessels (Lorenzi et al. 1985). Also, hyperglycaemia alters mRNA expression of integrin receptor subunits, adhesion molecules, thrombospondin-1 and vascular endothelial growth factor (VEGF), which are important for normal interactions of VEC with the extracellular matrix substratum in blood vessels (Stenina 2005).

Endothelial cell (EC) dysfunction in diabetes has also been attributed to the deleterious effects of free radicals (Ceriello 2003; Nishikawa et al. 2000). Numerous studies, both in diabetic patients and animal models of diabetes, have clearly shown that diabetes creates an oxidative environment *in vivo* (elevated levels of oxidative stress markers, on one hand, and subnormal levels of low-molecular weight anti-oxidants, on the other hand) (Cowell and Russell 2004; Dandona et al. 2005; Thornalley et al. 1996; Vincent et al. 2004) (see chater 3 for revision). Taking this into account, it has been suggested that the preservation of the anti-oxidant defence in cells and/or anti-oxidant therapies may ameliorate high glucose-induced complications (Tsuneki et al. 2007).

Despite all the evidence linking glucose toxicity to an increased risk of cardiovascular diseases, most studies have focused on the mechanism by which elevated intracellular glucose induces ED (Lee et al. 2002). So, very little is known about the regulation of glucose uptake in the cells of the vascular wall (Mann et al. 2003).

7.4 Glucose Metabolism in Endothelial and Smooth Muscle Cells

Glucose is actively metabolized in EC (Gerritsen and Burke 1985) and sustains both anaerobic and aerobic metabolism (Kreutzfeldt et al. 1990; Mertens et al. 1990). In rat coronary microvascular endothelial cells, >98% of incorporated glucose is metabolized to lactate (Kreutzfeldt et al. 1990). At physiological concentrations of glucose, the contribution of the hexose monophosphate pathway accounts for ±1.2% of glucose metabolism and the Krebs cycle for only ±0.04%, suggesting that in microvascular EC almost all of the energy obtained from catabolism of glucose is generated glycolytically. At lower glucose concentrations (±1 mM), oxidation of glucose via the Krebs cycle is higher. Thus, oxidative metabolism in EC is inhibited at physiological concentrations of glucose (a phenomenon known as the Crabtree effect) (Kreutzfeldt et al. 1990). Evidence obtained in human umbilical vein endothelial cells (HUVEC) suggests that fatty acids can also serve as an energy fuel, although oxidation of fatty acids accounts for only ±25% of the calculated ATP production in cells incubated with 5 mM glucose (Dagher et al. 2001).

In summary, EC synthesize ATP primarily via glycolysis, with a relatively low O_2 consumption (Culic et al. 1997; Dobrina and Rossi 1983; Mertens et al. 1990). Due to their high glycolytic activity, and also to their low energy demand, endothelial cells are able to withstand prolonged periods of substrate deprivation and can adapt to hypoxia (Culic et al. 1999a, b; Mertens et al. 1990).

Vascular smooth muscle cells (VSMC) also have a high rate of glycolysis, relying to a large extent on glycolytically generated ATP to sustain a variety of cell functions. Other than glucose, VSMC utilize several different substrates including short- or medium-chain fatty acids such as acetate and octanoate. Vascular smooth muscle metabolism and the influence of contraction on the metabolic fate of glucose and fatty acids have been extensively studied (Allen and Hardin 2000; Barron et al. 1991, 1994, 1998; Hardin and Paul 1995; Hardin and Roberts 1997).

7.5 Glucose Transporters

The entry of glucose into cells is a crucial step in metabolism, since glucose is a key fuel in mammals and an important metabolic substrate. Additionally, in mammalian cells, a tight regulation of blood glucose levels is needed to meet the energetic demands of the brain, a tissue that uses glucose as its primary energy source. So, adequate glucose flux into tissues provides maintenance of glucose homeostasis that is critical for health (Gorovits and Charron 2003; Wood and Trayhurn 2003).

The transfer of glucose across plasma membranes occurs via integral transport proteins. These transporters comprise two structurally and functionally distinct groups, whose members have been identified over the past two decades, namely: (1) the Na^+-dependent glucose co-transporters (SGLT, members of a large family of Na^+-dependent transporters, gene name SLC5A), mainly expressed in the apical membrane of renal and intestinal absorptive epithelial cells, that transport glucose against its concentration gradient and use ATP, and (2) the facilitative Na^+-independent sugar transporters (GLUT family, gene name SLC2A), expressed in all cells and that transport glucose down a concentration gradient (Gorovits and Charron 2003; Wood and Trayhurn 2003; Wright et al. 2007).

The facilitative transporters (GLUT) utilise the diffusion gradient of glucose (and other sugars) across plasma membranes and exhibit different substrate specificities, kinetic properties and tissue expression profiles (Table 7.1). GLUT transporters are predicted to have twelve membrane-spanning regions with intracellular located amino- and carboxyl-termini. To date, at least 12 facilitative glucose transporters have been cloned (Wood and Trayhurn 2003; Gorovits and Charron 2003).

GLUT1 is a high affinity (K_m = 1–2 mM, Table 7.1) glucose transporter that is ubiquitously expressed in most mammalian tissues. It provides basal glucose transport and, most importantly, transport of glucose through the blood-brain barrier, erythrocytes, and neuronal cell membranes.

The insulin-responsive glucose transporter, GLUT4, is found in heart, skeletal muscle and adipose tissue, being responsible for the reduction in the post-prandial rise in plasma glucose levels; it is also found in the brain (Wood and Trayhurn 2003; Gorovits and Charron 2003). In these tissues, the majority of GLUT4 molecules ($\pm$90%) are sequestered in intracellular vesicles in the absence of insulin or other stimuli such as muscle contraction (Shepherd and Kahn 1999; Kandror and Pilch 1996). Insulin acts by stimulating the translocation of specific GLUT4-containing vesicles from intracellular stores to the plasma membrane, resulting in an immediate 10–20-fold increase in glucose transport (Shepherd and Kahn 1999; Bryant et al. 2002). Gene regulation of GLUT4 expression is now recognized as an essential process in the modulation of glucose transport, particularly in diabetes and hypoxia (O'Brien and Granner 1996).

7.6 Glucose Transporters in the Endothelium and Vascular Smooth Muscle Cells

The distribution of GLUT transporters in mammalian cells is widespread, including EC from peripheral blood vessels and the blood-brain barrier (see Table 7.2). The high capacity of glucose transport into microvessels occurs primarily via the high-affinity, high-capacity GLUT1. However, other GLUT family members have been described in EC. GLUT3 mRNA and protein have been detected in the endothelium of human intraplacental microvessels, where it was proposed to play a potential role together with GLUT1 in sustaining glucose supply to the developing fetus (Hauguel-de-Mouzon et al. 1997). Also, GLUT4 has been detected in very low abundance in the rat forebrain microvasculature (McCall et al. 1997), but there is no evidence of GLUT4 activity in cerebral or retinal vasculature. Moreover, GLUT5 has been detected in the brain microvasculature (Mantych et al. 1993a, b; Takata et al. 1997), but there is no convincing evidence implicating a transport role for

Table 7.1 The family of facilitative glucose transporters (GLUT)

Transporter	No. amino acid residues	Main tissue localization	Insulin-sensitivity	Substrate and affinity	Function	Primary Rrefs.
GLUT1	492	Erythrocytes, blood-brain barrier, neurons, ubiquitous	No	Glucose (1—2 mM)	Basal glucose uptake	Mueckler et al. 1985;
GLUT2	524	Liver, kidney and intestinal epithelium, pancreatic ß-cells	No	Glucose (15—20 mM); fructose	Glucose sensing and transport by liver and pancreas	Fukumoto et al. 1988;
GLUT3	496	Neurons	No	Glucose (1—2 mM)	Basal glucose uptake	Kayano et al. 1988;
GLUT4	509	Skeletal muscle, heart, white and brown adipose cells	Yes	Glucose (5 mM)	Insulin-stimulated glucose uptake	Fukumoto et al. 1989;
GLUT5	501	Brush-border and basolateral membranes of intestinal epithelium, kidney, testes	No	Fructose (10—13 mM); glucose (very low affinity)	Transport from intestinal lumen into the cells and into sperm	Kayano et al. 1990;
GLUT6	507	Brain, spleen, leukocytes	No	Glucose	Not known	Doege et al. 2000a;
GLUT7	Not known	n.d.	n.d.	n.d.	Not known	Joost and Thorens 2001
GLUT8	477	Testis, brain, liver	No (yes in blastocytes)	Glucose (2 mM)	Not known	Carayannopoulos et al. 2000;
GLUT9	540	Liver, kidney	n.d.	n.d.	Not known	Phay et al. 2000
GLUT10	541	Liver, pancreas	No	Glucose (0.3 mM)	Not known	Dawson et al. 2001; McVie-Wylie et al. 2001
GLUT11	496	Heart, skeletal muscle	No	Glucose (low affinity); fructose	Not known	Doege et al. 2001; Wu et al. 2002; Sasaki et al. 2001
GLUT12	617	Heart, prostate, skeletal muscle, small intestine, white adipose tissue	Yes	n.d.	Not known	Rogers et al. 2002

Adapted from Gorovits and Charron 2003; Wood and Trayhurn 2003. n.d.-not determined.

Table 7.2 Glucose transporter isoforms detected in blood-brain, blood-retinal, and peripheral vascular endothelium

Isoform	Detection	Endothelial cell type	Main findings
		Blood-brain or blood-retinal barrier endothelium	
GLUT1	Protein	Human brain	Cerebral endothelium
	Protein	Human cerebellum microvessels	Cerebral endothelium and pericytes
	Protein	Human blood-brain barrier	Levels increased in hemangioblastoma
	Protein	Human blood-brain barrier	Decreased in Alzheimer's subjects
	Protein	Human blood-brain barrier	Decreased in Alzheimer's subjects
	Protein/mRNA	Human brain tumor vessels	More GLUT3 than GLUT1
	Protein	Human fetal brain	GLUT1 in 21-week endothelium
	mRNA	Bovine blood-brain barrier	Regulated by PKC activity
	mRNA	Bovine blood-brain barrier	↓ Glucose ↑ GLUT1 mRNA
	Protein/mRNA	Human retinal endothelial cells	mRNA for GLUT1 ↑ after 8 h Exposure to 15 mM glucose
	Protein/mRNA	Human retinal endothelial cells	Protein/mRNA unchanged after 4 to 72-h exposure to 22 mM D-glucose attenuated by microtubule inhibitor
	Protein	Human retinal capillary endothelium	Neovascular endothelium of proliferative retinopathy did not stain for GLUT1
	Protein	Bovine microvessel endothelial cells	Increased by glucose starvation
	Protein	Bovine large cortical endothelial cells	Detected
	Protein	Bovine retinal capillary	Upregulated by hypoxia via adenosine A_2 receptor and cAMP-PKA pathway
	Protein/mRNA	Bovine retinal endothelial cells	VEGF (50 ng/ml, 24 h) ↑ 3-*O*-methylglucose uptake via PKCβ
	Protein/mRNA	Rat brain	Asymmetric distribution: ~4-fold higher on abluminal membranes
	Protein	Rat brain	Streptozotocin-diabetes ↓ GLUT1 in retinal but not cerebral endothelium
	Protein	Rat brain capillary	Levels increased by 17β-estradiol
	Protein/mRNA	Rat blood-brain barrier	Levels increased by 17β-estradiol
	Protein	Rat brain capillary endothelial cells	Levels increased by glucose deprivation
	Protein	Rat blood-brain barrier	Present at birth already
	Protein	Rat blood-brain barrier	GLUT1 correlated with 2-DG uptake
	Protein	Rat cerebral microvessels	Hypoxia increased level
	Protein/mRNA	Rabbit blood-brain barrier	Regulated postnatal
	Protein	Rat retinal microvessels	Expressed at both luminal and abluminal endothelial cell membranes
	Protein	Mouse brain	Similar in fed and fasted animals
	Protein/mRNA	Mouse brain	Regulated postnatal
	Protein	Mouse brain	Asymmetric distribution: ~3-fold higher on abluminal membrane, similar density cortex, hippocampus, and cerebellum

Table 7.2 (continued)

Isoform	Detection	Endothelial cell type	Main findings
GLUT3	Protein/mRNA	Human placental microvessels	Absent in trophoblasts
	Protein/mRNA	Human brain tumor vessels	More GLUT3 than GLUT1
GLUT4	Protein/mRNA	Rat forebrain microvessels	Only detectable with high-stringency hybridization, low abundance
	Protein/mRNA	Bovine brain endothelial cells	Low and not in brain pericytes
GLUT5	Protein	Human brain endothelium	Cerebral cortical microvasculature but no transport function
		Endothelium from peripheral vascular beds	
GLUT1	Protein	Rat adrenal capillary and aorta	Glucose deprivation ↑ 2-deoxyglucose uptake and GLUT1 expression ~12 h sooner than in brain endothelial cells
	Protein	Chicken retina	
	Protein/mRNA	Rat heart	GLUT4 not detected
	Protein/mRNA	Human retina	Diffuse reactivity throughout retina
	Protein/mRNA	Monkey retina	
	mRNA	Human retinal endothelial cells	GLUT1 maximal at 15 mM glucose
	Protein	Fetal endothelium, human placenta	Detected in fetal but not maternal vessels
	Protein	Fetal endothelium, marmoset placenta	
	Protein	Fetal endothelium, rat placenta	
	Protein	Fetal endothelium, human placenta	
	Protein	Human iris capillary	Absent in diabetes
	Protein	Human cornea	Absent in diabetes
	Protein	Human testis microvessels	
	Protein	Human fetal vein endothelium	Short-term insulin no effect

PKC, protein kinase C; PKA, protein kinase A; VEGF, vascular epidermal growth factor, 2-DG, 2-deoxyglucose.
Taken from Mann et al. (2003). Used with permission.

GLUT5 in the blood-brain or blood-retinal barriers. Finally, there is no evidence that GLUT6, GLUT8 or GLUT11 are expressed or have a functional role in VEC or VSMC (Mann et al. 2003).

It was recently reported the presence and asymmetric distribution of glucose transporters GLUT-1 to -5 and SGLT-1 in the endothelium of rat coronary, cerebral, renal, and mesenteric arteries (Gaudreault et al. 2004, 2006). The subcellular localization of these transporters (predominantly located on the abluminal side of the endothelium) may facilitate trans-endothelial transport of glucose in small contractile arteries (Gaudreault et al. 2006, 2008). The presence of GLUT-1 to 5 and SGLT-1 was also detected in cultures of human coronary artery endothelial cells (HCAECs), although their subcellular distribution lacks the luminal/abluminal asymmetry observed in intact endothelium (Gaudreault et al. 2008). Interestingly, significant differences in the subcellular localization of an incorporated fluorescent glucose analog (2-NBDG) between the endothelium of rat coronary arteries and HCAECs were found, and this mirrored the distributions of their glucose transporters (Gaudreault et al. 2008).

7.7 Glucose Transport in the Endothelium and the Metabolic Syndrome

Diabetes mellitus (DM) and the MS are associated with an increased risk of cardiovascular disease. Substantial clinical and experimental evidence suggest that both diabetes and insulin resistance cause a combination of EDs, which may diminish the anti-atherogenic role of the vascular endothelium.

Hyperglycaemia has been implicated in the pathogenesis of micro- and macrovascular complications of diabetes. In recent years, the role of chronic hyperglycaemia in the development of the microvascular complications seen in diabetes and insulin resistance has been clearly established. However, the biochemical or cellular links between elevated blood glucose levels and the vascular functional and structural pathological changes remain incompletely understood (Hadi and Suwaidi 2007). In this context, knowledge concerning the mechanisms involved in glucose uptake by EC, and their regulation, seems important, because changes in glucose uptake by these cells under these conditions are to be expected (Machado et al. 2006). However, few studies have been performed on this subject.

The effects of elevated glucose on EC metabolism include: (1) a reduction in the cytosol size of EC due to the accumulation of NADH and transformation of pyruvate to lactate (Dobrina and Rossi 1983; Hingorani and Brecher 1987); (2) an activation of the polyol pathway (Cohen 1993; Taylor and Agius 1988); and (3) an increase in the generation of superoxide anions (which react with NO to form peroxynitrite, which upon decomposition generate a strong oxidant with reactivity similar to hydroxyl radicals) (Beckman 1996). Interestingly, human EC exposed to hyperglycaemia in established DM are more sensitive to ROS, since intracellular levels of glutathione, vitamin E, superoxide dismutase, catalase and ascorbic acid are significantly decreased (reviewed in Droge 2002; Halliwell 1993).

7.7.1 Effect of Hyperglycaemia/Diabetes Mellitus

At the blood-brain barrier (BBB) level, the passage of glucose across the EC is mainly mediated by GLUT1 (Gerhart et al. 1989; Pardridge et al. 1990a). The general consensus is that GLUT1 is asymmetrically distributed between the luminal and abluminal membranes of the BBB (ratio 1:4, Farrell and Pardridge 1991). The importance of GLUT1 in the blood-brain passage of glucose is well demonstrated by the fact that children with GLUT1 deficiency exhibit impaired glucose transport across the BBB associated with infantile seizures and developmental delay (Klepper et al. 1999).

Recent studies have shown that the brain EC line RBE4, derived from primary cultures of rat brain capillary endothelium, also transports 3-O-methylglucose via a facilitative glucose transporter (apparently GLUT1) sensitive to glucose deprivation (Regina et al. 1997).

McCall et al. (1986) first demonstrated *in vivo* that chronic hypoglycaemia results in increased glucose transport to the brain of rats. Up-regulation of BBB GLUT1 protein and mRNA expression was shown to mediate this increased transport (Kumagai et al. 1995; Kumagai 1999). Similarly, an increase in glucose uptake from blood to brain following a few days of starvation or moderate hypoglycaemia was observed *in vivo* in humans (Blomqvist et al. 1991; Paulson and Hasselbalch 1997). Glucose starvation was also found to enhance glucose uptake in cultured brain, adrenal capillary, and aortic EC and to increase GLUT1 protein expression (Gaposchkin et al. 1996; Takakura et al. 1991).

On the other hand, chronic hyperglycaemia/diabetes has been associated with a reduction of BBB glucose transport and GLUT1 expression (Harik et al. 1988; Gjedde and Crone 1981; Pardridge et al. 1990b; McCall et al. 1982, 1984; Mooradian et al. 1991; Cornford et al. 1995; Pouliot and Beliveau 1995). Palmitoylation seems to be involved in the modulation of BBB glucose transporters in hyperglycaemia, as palmitoylation of GLUT1 was increased in hyperglycaemic and diabetic rats (Pouliot and Beliveau 1995). So, these studies suggest that prolonged hyperglycaemia *in vivo* decreases BBB glucose transport. However, other studies have reported either no change or even up-regulation of GLUT1 expression (Pelligrino et al. 1992; Jacob et al. 2002).

The central nervous system has not, traditionally, been thought of as a target of damage associated with diabetic hyperglycaemia. However, recent evidence suggests that the brain may indeed be damaged by such conditions. A report utilizing data from approximately 10,000 women showed a significant association between the presence of diabetes (and duration of disease) and impaired cognitive performance (Gregg et al. 2000). Also, the reduction in GLUT1 levels at the BBB may be involved in the central signals of hypoglycaemia in diabetics adapted to a hyperglycaemic state, and which suffer a sudden glycaemic drop, even if the glycaemic levels at the periphery are preserved.

The GLUT1 glucose transporter also mediates glucose entry into the EC of the inner human blood-retinal barrier (BRB) (Busik et al. 2002; Knott et al. 1996; Kumagai et al. 1994). As in the brain endothelium, GLUT1 is expressed on luminal and abluminal membranes of human and rat BRB.

In many cell types, exposure to high glucose concentrations or diabetes down-regulates GLUT1 (see above). In agreement with this general observation, GLUT1 appears to be absent in human iris and corneal capillaries in DM (Kumagai et al. 1994), and in the rat BRB, streptozotocin-induced diabetes apparently increases the K_m value for glucose transport (Ennis et al. 1982). However, other studies found quite opposite results. On the inner BRB in Goto-Kakizaki (GK) rats, an animal model of long-standing diabetes, there is no compensatory down-regulation of GLUT1 (Fernandes et al. 2003). Moreover, exposure of primary cultures of human retinal EC to elevated glucose concentration (22 mM) increases the V_{max} 2.5-fold with no changes in GLUT1 mRNA or protein levels (Busik et al. 2002). Also, Knott et al. (1996) reported that GLUT1 mRNA levels increased in human retinal EC after exposure for 8 h to 15 mM glucose (although it is worth noting that mRNA levels were similar in cells cultured in 5 or 25 mM glucose). Finally, Mandarino et al. (1994) showed that high glucose down-regulates glucose transport and GLUT1 protein in bovine retinal pericytes, but not in EC. So, a localized up-regulation of GLUT1 expression at the luminal surface of BRB in long-standing diabetics not affected by retinopathy may be associated with the deleterious effects of chronic hyperglycaemia on the retinal microvasculature.

The effect of high glucose concentration upon glucose uptake by bovine aortic EC has been studied by a number of groups, and the results obtained have been rather contradictory. In some studies, exposure of bovine aortic EC to elevated glucose had negligible effects on 2-deoxyglucose or 3-O-methylglucose transport or GLUT1 mRNA or protein levels, whereas it decreased the V_{max} for transport in bovine and human aortic SMC (Kaiser et al. 1993; Viñals et al. 1999). It was suggested that the insensitivity of glucose transport in EC to hyperglycaemia, and the differential regulation of glucose transport in primary cultures of SMC and EC might contribute to the ED associated with chronic hyperglycaemia.

In contrast, other studies have shown that bovine aortic EC and SMC down-regulate the rate of glucose transport, the mRNA and the protein content of their typical glucose transporter (GLUT1) as well as its plasma-membrane abundance, in the face of hyperglycaemia (Alpert et al. 2002, 2005; Sasson et al. 1996; Totary-Jain et al. 2005). It was suggested that this mechanism might provide protection against the deleterious effects of increased intracellular glucose levels.

Finally, an increase in glucose transport in a bovine EC line (GM7373) was observed with a high glucose concentration. However, no changes in GLUT1 expression were detected (Giardino et al. 1994). So, this particular EC line may well not reflect transport processes in primary bovine EC cultures.

The effect of hyperglycaemia upon glucose uptake has also been studied in the EC from the septal coronary artery of Wistar rats. Long-term hyperglycaemia, induced by streptozotocin, significantly down-regulated GLUT1, 3, 4 and 5 and dramatically up-regulated GLUT-2, leaving SGLT-1 unchanged. The authors concluded that the high susceptibility of EC to glucose toxicity may be the result of the sub-cellular organisation of their GLUTs and the increased expression of GLUT2 (Gaudreault et al. 2004).

In bovine cultured aortic EC and EMC, the antidiabetic drug metformin, used to restore insulin sensitivity in diabetic patients, caused a time- and dose-dependent increase in the rate of 2-deoxyglucose and 3-O-methylglucose uptake (Sasson et al. 1996). It was concluded that metformin increases translocation of GLUT1 to the plasma membrane rather than affecting its intrinsic activity, confirming similar findings in adipocytes and cardiac myocytes (Fischer et al. 1995; Matthaei et al. 1991).

7.7.2 Effect of Pro-Oxidants

In primary cultures of bovine aortic EC, hydrogen peroxide (25 μM) increased the rate of glucose uptake (Altman et al. 2004). Similarly, high levels of unconjugated bilirubin, which presents pro-oxidant properties in these cells (Kapitulnik 2004; Cohen et al. 2006), was found to increase the rate of glucose transport in these same cells either exposed to normal or high glucose concentrations. This effect was associated with an increase in the expression of GLUT1 (Cohen et al. 2006). Further support to the findings on oxidative stress-induced stimulation of glucose transport was obtained with 4-hydroxy tempol (TPL) in the same cells. This compound, when present at pro-oxidative concentrations, increased the rate of uptake of glucose in cells exposed to either normal or high concentration of glucose, an effect associated with an increase in GLUT1 mRNA and protein content (Alpert et al. 2004).

These observations in EC are in good contrast to what happens in skeletal muscle cells and adipocytes, in which ROS significantly reduce the expression of the insulin sensitive transporter (GLUT4; Pessler et al. 2001).

The results presented above may explain the critical role of the combination of hyperglycaemia and an intense oxidative stress in the development of EC dysfunction (Cohen et al. 2007). Moreover, these findings indicate that potent pro-oxidants augment the rate of glucose transport in VEC cultures by increasing the expression of GLUT1. Of interest is the observation that, although hyperglycaemia induced a moderate oxidative stress, it failed to induce a similar compensatory up-regulatory mechanism. On the contrary, the expression of GLUT1 in VEC and the rate of glucose uptake decreased (see above). So, it appears that the high glucose-induced auto-regulatory protective mechanism in bovine aortic EC collapses in the presence of other potent pro-oxidants.

The lack of effect of several anti-oxidants (aminoguanidine, N-acetylcysteine, trolox, and vitamin C) upon the diminished rate of glucose transport in VEC under high glucose conditions (Altman et al. 2004) suggests that glucose-derived free radicals are not involved in the down-regulation of glucose transport in these cells under high glucose conditions. These results further support the conclusion on a functional dissociation between high glucose-induced oxidative stress and high glucose-induced down-regulation of glucose transport system in VEC.

In conclusion, two processes occur simultaneously in VEC cultures exposed to high glucose levels: first, an augmented production of ROS due to metabolic effects and non-enzymatic glycation of glucose, and second, the down-regulation of glucose transport due to destabilization of GLUT1 mRNA. Pro-oxidants also induce oxidative stress in VEC, but differently from hyperglycaemia, they up-regulate the glucose transport system. This disparity in the mode of regulation of the glucose transport system by two oxidative stressful conditions is intriguing. Cohen et al. (2007) proposed a model for pro-oxidants-induced collapse of the natural protective mechanism against hyperglycaemia in bovine aortic EC primary cultures, leading to EC dysfunction (Fig. 7.1). These authors suggest that when hyperglycaemia in diabetic patients is complicated with other pro-oxidative challenges, the EC monolayer in blood vessels loses its first line of defence, namely down-regulation of glucose transport, and becomes vulnerable to detrimental oxidative

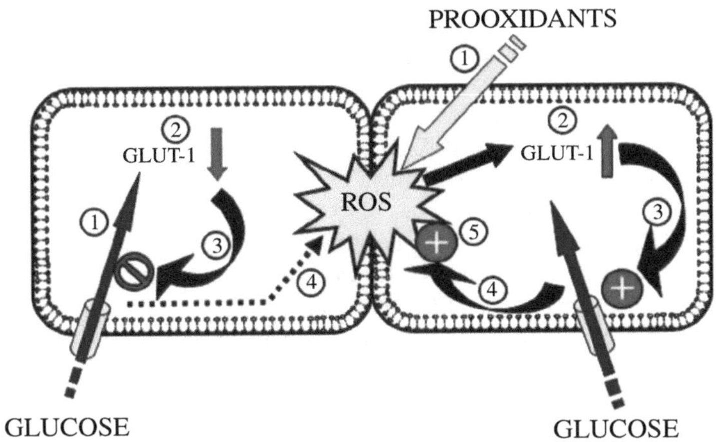

Fig. 7.1 A model for pro-oxidants-induced collapse of the natural protective mechanism against hyperglycaemia in bovine aortic endothelial cell primary cultures. Left: Long-term hyperglycaemia induces down-regulation of the glucose transport system following accumulation of glucose-derived intracellular metabolites/signals (1), which destabilize GLUT-1 mRNA and consequently reduce GLUT-1 content in the cell and in the plasma membrane (2), thus lowering the rate of glucose transport (3). This mechanism limits production of glucose-derived free radical and induces a weak oxidative stress. Right: Pro-oxidants enter the cell and induce an intense oxidative stress (1), which then operate a compensatory response to augment GLUT-1 expression (2). Consequently GLUT-1 abundance in the plasma membrane is increased and the rate of glucose transport is augmented (4) to produce more glucose-derived free radicals (5), and vice versa. Taken from Cohen et al. (2007). Used with permission

interactions. Among pathological conditions known to create an oxidative stress are conditions associated with the MS, such as hyperlipidaemia or hypertriglyceridaemia (Galle et al. 2006; Nitenberg et al. 2006; Rebolledo and Actis Dato 2005) and pro-oxidative function of various micro- and macronutrients (Dandona et al. 2005), dietary supplements, environmental pollutants and certain pharmaceuticals (Gonzalez-Flecha 2004; Parke and Sapota 1996; Tao et al. 2003). So, prevention of such harmful conditions and exposures in hyperglycaemic individuals may better preserve the auto-regulatory potential of VEC and delay or prevent EC dysfunction.

7.7.3 *Effect of Insulin and Angiogenic Mediators: Insulin-Like Growth Factor (IGF-1) and Vascular Endothelial Growth Factor (VEGF)*

It is still under debate whether glucose uptake and metabolism in EC are regulated by insulin. Early studies of the *in vivo* effects of insulin on glucose transfer across the BBB in humans and rats only noted small changes in glucose fluxes during infusion of insulin (Hertz et al. 1981; Namba et al. 1987). Also, insulin did not affect glucose uptake in human and bovine macrovascular (aortic, umbilical vein and pulmonary artery) EC (Gosmanov et al. 2006; Pekala et al. 1990; Parra et al. 1998;

Corkey et al. 1981; Bar et al. 1988). However, human aortic endothelium was reported to become insulin responsive after prolonged (pre)incubation in medium supplemented with high glucose (Gosmanov et al. 2006). Insulin, moreover, failed to affect glucose uptake in microvascular EC, such as bovine brain and retinal EC (Betz et al. 1973, 1983; McCall et al. 1997; Takakura et al. 1991). Finally, a recent study by Artwohl et al. (2007) showed that insulin failed to stimulate glucose transport in human micro- (HRECs) and macrovascular EC (HUVECs, HAVECs and HAECs).

However, other studies have shown a stimulating effect of insulin upon endothelial glucose uptake. Glucose transport was increased by insulin in bovine adipose tissue-derived microvessel EC (Bar et al. 1988). In bovine retinal EC, Allen et al. (1986) documented a time- and protein synthesis-dependent increase in glucose transport in response to insulin. In another study, insulin stimulated uptake of ^{3}H-2-deoxyglucose and ^{3}H-O-methyl-D-glucose in cultured rabbit coronary microvessel endothelium (Gerritsen and Burke 1985); however, this effect required both serum and glucose deprivation (Gerritsen et al. 1988). Also, treatment of diabetic rats with insulin restored the brain uptake index (BUI) measurements for 3-O-methylglucose uptake, which were reduced in diabetic animals, to values of non-diabetic animals (Mooradian et al. 1991). Finally, uptake of the glucose analog (2-NBDG) in HCAECs and in EC of rat coronary artery was stimulated by insulin (Gaudreault et al. 2008).

Increased expression of the insulin-sensitive glucose transporter (GLUT4) was originally reported in microvascular EC after administration of insulin to animals *in vivo* (Vilaro et al. 1989), although a subsequent study was unable to reproduce these results (Slot et al. 1990).

Insulin-like growth factor-1 (IGF-1) was found to stimulate glucose transport in primary cultures of bovine retinal EC via phosphatidylinositol (PI) 3-kinase and PKC-dependent mechanisms, an effect thought to involve the translocation of cytosolic GLUT1 to the plasma membrane (Debosch et al. 2002). In these same cells, VEGF stimulated 3-O-methylglucose transport via a PKC-ß-mediated translocation of pre-existing cytosolic GLUT1 transporters to the plasma membrane (Sone et al. 2000). This action of IGF-1 and VEGF on GLUT1 is reminiscent of the actions of insulin in GLUT4-sensitive tissues. Moreover, VEGF caused an approximately threefold increase in 2-deoxyglucose uptake and a fivefold increase in GLUT1 mRNA levels in a bovine aortic EC clone (JVO17A) (Pekala et al. 1990).

IGF-1 and VEGF are two well-established angiogenic stimulating factors, playing primary roles in EC (see Chapter 5). Beside other stimuli, both these factors can be modulated by insulin. Their effect on glucose transport (enhancement) is, therefore, in total agreement with their well established survival role.

7.7.4 Effect of Inflammatory Mediators

Pro-inflammatory cytokines, such as tumour necrosis factor-α (TNF-α), interleukin (IL)-1ß, and interferon-γ (IFN-γ), are mediators of host responses to infection or inflammation. Pro-inflammatory cytokines induce a wide range of effects at the vascular level. ROS generated in endothelial and SMC in response to cytokines can function as intracellular signalling molecules, modulating permeability, leukocyte adhesion, actin filament organization, redox-sensitive transducers and transcription factors.

There is only one study on the effects of pro-inflammatory cytokines on glucose transport in EC. In that work, treatment of bovine aortic EC with TNF-α increased hexose transport and GLUT1 mRNA levels (Pan et al. 1995).

In cultured coronary microvascular EC, activation of H_1 receptors by acute histamine stimulates glucose transport, an effect thought to involve modulation of GLUT1 expression and/or activity (Thomas et al. 1995). On the other hand, transport of 3-O-methylglucose in HUVEC is unaffected by ATP, adenosine, or histamine (Parra et al. 1998), reflecting potential differences between micro- and macrovascular endothelium. Because histamine-mediated release of PGI_2 from human macrovascular and microvascular EC is impaired by hyperglycaemia (Sobrevia and Mann 1997), further studies of the effects of elevated glucose in coronary microvascular EC are warranted.

There are no reports on the acute or chronic effects of NO on glucose transport in cultured VEC. However, a role for NO as a stimulator of glucose transport in skeletal muscle (which is GLUT4-mediated) has been documented (Etgen et al. 1997). In contrast, myocardial glucose uptake is increased in the presence of the nitric oxide synthase (NOS) inhibitor L-NAME and in preparations isolated from endothelial NOS (eNOS) knock-out mice (Tada et al. 2000). Considering that in diseases such as ischemic heart disease, DM, hypertension, and hypercholesterolaemia, NO synthesis is altered and substrate utilization changes, and that NO-dependent mechanism(s) play an important role at sites of inflammation, further studies on this subject are warranted.

7.7.5 *Effect of Hypoxia*

Retinal hypoxia often precedes proliferative diabetic retinopathy, and an increase in intracellular glucose in retinal vascular cells is thought to be an important factor in the development of diabetic retinopathy (Kumagai 1999). Takagi et al. (1998) investigated the effects of hypoxia on GLUT1 mRNA expression in bovine cultured retinal capillary EC. Exposure of EC to hypoxia caused time-dependent changes in GLUT1 mRNA levels, with an 8.9-fold increase detected after 12 h paralleled by a 2- to 3-fold increase in 2-deoxyglucose transport and immunoreactive GLUT1. The hypoxia-mediated increase in GLUT1 transport activity was mediated in part via adenosine A_2 receptors and the cAMP-PKA pathway, since antagonists of A_2 purinoceptors and PKA suppressed hypoxia-induced GLUT1 expression (Takagi et al. 1998). Up-regulation of glucose transport may have important implications for microvascular dysfunction in diabetic retinopathy (Kumagai 1999). Increased translocation of GLUT1 transporters to the plasma membrane of retinal endothelial cells in response to hypoxia and growth factors would lead to increased glucose transport, resulting ultimately in metabolic and structural changes in the BRB. In view of the sensitivity of the retinal vasculature to hyperglycaemia-mediated injury, it is surprising how little is known about the cellular mechanisms regulating glucose transport and metabolism at this level.

Hypoxia also modulates glucose transport in human fetal and bovine aortic EC (Loike et al. 1992). Endothelial cells cultured under low oxygen conditions (14 mmHg) for up to 96 h exhibited increased rates of glucose transport and generated more lactic acid than normoxic cells. Moreover, activation of glucose transport by hypoxia required several hours and was associated with an increased expression of GLUT1 protein and mRNA. As inhibitors of oxidative phosphorylation mimicked the effects of hypoxia, it was proposed that oxidative metabolism might serve as an important signal for adaptive responses in EC to hypoxia (Loike et al. 1992).

7.8 Glucose Transport in the Vascular Smooth Muscle and the Metabolic Syndrome

7.8.1 Effect of Diabetes/Hypertension

The insulin-responsive glucose transporter GLUT4 is expressed in vascular smooth muscle (Banz et al. 1996; Brosius et al. 1992; Cooper et al. 1993; Kahn et al. 1995; Standley and Rose 1994; Bryant et al. 2002). In arterial VSMC, GLUT4 participates in constitutive, non-insulin-dependent glucose uptake (Atkins et al. 2001; Park et al. 2005). This unusual property distinguishes VSMC from other tissues that express GLUT4 since in those tissues GLUT4 largely resides in intracellular vesicles until translocated to the plasma membrane in response to insulin or other physiological stimuli (Bryant et al. 2002). Although GLUT4 is the main glucose transporter present in VSMC, these cells have been shown to express GLUT1 as well (Atkins et al. 2001).

Several works have demonstrated that in vascular diseases caused by diabetes and hypertension, GLUT4 expression in VSMC is decreased, with a concomitant decrease in basal vascular glucose uptake (Atkins et al. 2001, 2005; Marcus et al. 1994; Park et al. 2005). GLUT4 is necessary for agonist-induced contraction (Park et al. 2005), but chronic GLUT4 knock-out was associated with increased vascular reactivity compared with that in wild-type mice, suggesting that chronic absence or reduction of GLUT4 expression in VSMC leads to opposite effects observed with acute inhibition of GLUT4 (Park et al. 2005). So, GLUT4 appears to be an important glucose transporter in VSMC and changes in GLUT4 expression in these cells may account for some of the contractile abnormalities associated with vascular diseases like hypertension. In line with this conclusion, it was recently demonstrated that preservation of GLUT4 expression (by using GLUT4 transgenic mice) in mouse aorta prevents the enhanced arterial reactivity observed in hypertension, possibly via an effect on myosin phosphatase activity (Atkins et al. 2007). Moreover, treatment of DOCA salt-hypertensive rats with troglitazone and rosiglitazone, which are activators of peroxisome proliferator-activated receptor-gamma (PPAR-gamma), a stabilizer of GLUT4 expression, caused an increase in the expression levels of GLUT4 and a decrease in systolic blood pressure, an effect that appears to be mediated through activation of PI 3-kinase/Akt (Atkins et al. 2005). Experimental DM was also associated with a down-regulation of GLUT4 expression and 2-deoxyglucose uptake in renal VSMC, perhaps contributing to glomerular hyperfiltration and hypertension in the early stages of diabetes (Marcus et al. 1994).

Hyperglycaemia has also been shown to affect VSMC GLUT1-mediated glucose uptake. Elevating glucose from 1.2 to 22 mM (24 h) decreased GLUT1 protein levels and the V_{max} for 2-deoxyglucose transport, although GLUT1 mRNA levels were paradoxically unaffected by changes in glucose concentrations (Kaiser et al. 1993). Similar findings have been reported in quiescent VSMC from rat aorta, where elevated glucose decreased transport rates for 2-deoxyglucose and glucose and GLUT1 transporter protein levels (Howard 1996; Quinn and McCumbee 1998). Because intracellular glucose concentrations remained elevated in cells exposed to 20 mM glucose, Howard (1996) concluded that down-regulation of GLUT1 transporter activity does not appear to normalize intracellular glucose levels in VSMC and may account for the toxicity of elevated glucose in DM.

7.8.2 Effect of Insulin and Platelet-Derived Growth Factor

Insulin stimulates glucose transport in rat aortic SMC and A10 and A7r5 VSMC lines, and treatment of A7r5 cells with insulin or IGF-I increased glucose transport activity (Standley and Rose 1994). Interestingly, the stimulatory actions of insulin on glucose transport in rat aortic SMC were attenuated in cells pre-adapted to 25 mM D-glucose (Fujiwara and Nakai 1996).

Anti-diabetic agents such as troglitazone and metformin, used to enhance insulin sensitivity in humans, have been reported to increase hexose transport and GLUT1 mRNA levels in human and bovine aortic SMC, respectively (Kihara et al. 1998; Sasson et al. 1996).

Treatment of rat aortic smooth muscle cells with platelet-derived growth factor (PDGF) results in a stimulation of 2-deoxyglucose transport (MacKenzie et al. 2001). Although GLUT1 is the predominant isoform in this cell type, this study did not establish whether stimulation of transport involved an intrinsic activation of existing carriers or a redistribution of GLUT1 transporters. PDGF is crucial for SMC homeostasis, promoting proliferation and survival of these cells, and attachment to angiogenic vessels, resulting in vascular maturation. Its effect on glucose uptake might be related with SMC survival.

7.9 Conclusions

The significant advances in the molecular biology of glucose transporters and intracellular signalling pathways within the last decade provide the necessary tools for characterizing the molecular and functional regulation of nutrient transporters in EC and SMC derived from the brain, retinal, and peripheral vasculature in health and disease.

Endothelium-dependent vascular relaxation is markedly impaired in diseases such as DM, atherosclerosis, hypertension, and preeclampsia. Because disease-induced alterations in plasma levels of glucose and insulin modulate vascular relaxation, knowledge on the regulation of transport and metabolism of glucose in vascular EC and SMC appears as a most important target.

References

Allen LA, Gerritsen ME. *Regulation of hexose transport in cultured bovine retinal microvessel endothelium by insulin.* Exp Eye Res. 1986; 43: 679–86.

Allen TJ, Hardin CD. *Influence of glycogen storage on vascular smooth muscle metabolism.* Am J Physiol Heart Circ Physiol. 2000; 278: H1993–H2002.

Alpert E, Gruzman A, Totary H, Kaiser N, Reich R, Sasson S. *A natural protective mechanism against hyperglycaemia in vascular endothelial and smooth-muscle cells: role of glucose and 12-hydroxyeicosatetraenoic acid.* Biochem J. 2002; 362: 413–22.

Alpert E, Altman H, Totary H, Gruzman A, Barnea D, Barash V, Sasson S. *4-Hydroxy tempol-induced impairment of mitochondrial function and augmentation of glucose transport in vascular endothelial and smooth muscle cells.* Biochem Pharmacol. 2004; 67: 1985–95.

Alpert E, Gruzman A, Riahi Y, Blejter R, Aharoni P, Weisinger G, Eckel J, Kaiser N, Sasson S. *Delayed autoregulation of glucose transport in vascular endothelial cells.* Diabetologia. 2005; 48: 752–5.

Altman H, Alpert E, Sasson S. *Do glucose-derived reactive oxygen species contribute to the autoregulation of glucose transport in vascular endothelial and smooth muscle cell?* In: Simionecu M, Sima A, Popov D (eds) Cellular dysfunction in atherosclerosis and diabetes: reports from bench to bedside. Romanian Academy Publishing House, Bucharest. 2004; 274–82

Artwohl M, Brunmair B, Fürnsinn C, Hölzenbein T, Rainer G, Freudenthaler A, Porod EM, Huttary N, Baumgartner-Parzer SM. *Insulin does not regulate glucose transport and metabolism in human endothelium.* Eur J Clin Invest. 2007; 37, 643–50.

Atkins KB, Johns D, Watts S, Clinton Webb R, Brosius FC3. *Decreased vascular glucose transporter expression and glucose uptake in DOCA-salt hypertension.* J Hyperten. 2001; 19: 1581–7.

Atkins KB, Northcott CA, Watts SW, Brosius FC. *Effects of PPAR-gamma ligands on vascular smooth muscle marker expression in hypertensive and normal arteries.* Am J Physiol Heart Circ Physiol. 2005; 288: H235–43.

Atkins KB, Prezkop A, Park JL, Saha J, Duquaine D, Charron MJ, Olson AL, Brosius FC 3rd. *Preserved expression of GLUT4 prevents enhanced agonist-induced vascular reactivity and MYPT1 phosphorylation in hypertensive mouse aorta.* Am J Physiol Heart Circ Physiol. 2007; 293: H402–8.

Banz WJ, Abel MA, Zemel MB. *Insulin regulation of vascular smooth muscle glucose transport in insulin-sensitive and resistant rats.* Horm Metab Res. 1996; 28: 271–5.

Bar RS, Siddle K, Dolash S, Boes M, Dake B. *Actions of insulin and insulinlike growth factors I and II in cultured microvessel endothelial cells from bovine adipose tissue.* Metabolism. 1988; 37: 714–20.

Barron JT, Koop SJ, Tow JP, Parrillo JE. *Differential effects of fatty acids on glycolysis and glycogen metabolism in vascular smooth muscle.* Biochim Biophys Acta. 1991; 1093: 125–34.

Barron JT, Koop SJ, Tow JP, Parrillo JE. *Fatty acid, tricarboxylic and cyclic metabolites and energy metabolism in vascular smooth muscle.* Am J Physiol Heart Circ Physiol. 1994; 267: H764–9.

Barron JT, Barany M, Gu L, Parrillo JE. *Metabolic fate of glucose in vascular smooth muscle during contraction induced by noradrenaline.* J Mol Cell Cardiol. 1998; 30: 709–19.
Beckman JS. *Oxidative damage and tyrosine nitration from peroxynitrite.* Chem Res Toxicol. 1996; 9: 836–44.
Betz AL, Gilboe DD, Yudilevich DL, Drewes LR. *Kinetics of unidirectional glucose transport into the isolated dog brain.* Am J Physiol. 1973; 225: 586–92.
Betz AL, Bowman PD, Goldstein GW. *Hexose transport in microvascular endothelial cells cultured from bovine retina.* Exp Eye Res. 1983; 36: 269–77.
Blomqvist G, Gjedde A, Gutniak M, Grill V, Widen L, Stoneelander S, Hellstrand E. *Facilitated transport of glucose from blood to brain in man and the effect of moderate hypoglycemia on cerebral glucose utilization.* Eur J Nuclear Med. 1991; 18: 834–7.
Brosius FC, Briggs JP, Marcus RG, Barac-Nieto M, Charron MJ. *Insulin-responsive glucose transporter expression in renal microvessels and glomeruli.* Kidney Int. 1992; 42: 1086–92.
Bryant NJ, Govers R, James DE. *Regulated transport of the glucose transporter GLUT4.* Nature Rev Mol Cell Biol. 2002; 3: 267–77.
Busik JV, Olson LK, Grant MB, Henry DN. *Glucose-induced activation of glucose uptake in cells from the inner and outer blood-retinal barrier.* Invest Ophthalmol Vis Sci. 2002; 43: 2356–63.
Ceriello A. *New insights on oxidative stress and diabetic complications may lead to a 'causal' anti-oxidant therapy.* Diabetes Care. 2003; 26: 1589–96.
Cohen RA. *Dysfunction of vascular endothelium in diabetes mellitus.* Circulation. 1993; 87: V67–76.
Cohen G, Livovsky DM, Kapitulnik J, Sasson S. *Bilirubin increases the expression of glucose transporter-1 and the rate of glucose uptake in vascular endothelial cells.* Rev Diabet Stud. 2006; 3: 127–33.
Cohen G, Riahi Y, Alpert E, Gruzman A, Sasson S. *The roles of hyperglycaemia and oxidative stress in the rise and collapse of the natural protective mechanism against vascular endothelial cell dysfunction in diabetes.* Arch Physiol Biochem. 2007; 113: 259–67.
Cooper DR, Khalakdina A, Watson JE. *Chronic effects of glucose on insulin signaling in A-10 vascular smooth muscle cells.* Arch Biochem Biophys. 1993; 302: 490–498.
Corkey RF, Corkey BE, Gimbrone MA. *Hexose transport in normal and SV40-transformed human endothelial cells in culture.* J Cell Physiol. 1981; 106: 425–34.
Cornford EM, Hyman S, Cornford ME, Clare-Salzler M. *Down-regulation of blood-brain glucose transport in the hyperglycemic nonobese diabetic mouse.* Neurochem Res. 1995; 20: 869–73.
Cowell RM, Russell JW. *Nitrosative injury and anti-oxidant therapy in the management of diabetic neuropathy.* J Investig Med. 2004; 52: 33–44.
Culic O, Decking UKM, Bergschneider E, Schrader J. *Purinogen is not an endogenous substrate used in endothelial cells during substrate depletion.* Biochem J. 1999a; 338: 523–7.
Culic O, Decking UKM, Schrader J. *Metabolic adaptation of endothelial cells to substrate deprivation.* Am J Physiol Cell Physiol. 1999b; 276: C1061–68.
Culic O, Gruwel MLH, Schrader J. *Energy turnover of vascular endothelial cells.* Am J Physiol Cell Physiol. 1997; 273: C205–13.
Dagher Z, Ruderman N, Tornheim K, Ido Y. *Acute regulation of fatty acid oxidation and AMP-mediated protein kinase in human umbilical vein endothelial cells.* Circ Res. 2001; 88: 1276–82.
Dandona P, Aljada A, Chaudhuri A, Mohanty P, Garg R. *Metabolic syndrome: a comprehensive perspective based on interactions between obesity, diabetes, and inflammation.* Circulation. 2005; 111: 1448–54.
Debosch BJ, Deo K, Kumagai AK. *Insulin-like growth factor-1 effects on bovine retinal endothelial cell glucose transport: role MAP kinase.* J Neurochem. 2002; 81: 728–34.
Dobrina A, Rossi F. *Metabolic properties of freshly isolated bovine endothelial-cells.* Biochim Biophys Acta. 1983; 762: 295–301.
Droge W. *Free radicals in the physiological control of cell function.* Physiol Rev. 2002; 82: 47–95.

Ennis SR, Johnson JE, Pautler EL. *In situ kinetics of glucose transport across the blood-retinal barrier in normal rats and rats with streptozocin-induced diabetes.* Invest Ophthalmol Vis Sci. 1982; 23: 447–56.

Etgen GJ, Fryburg DA, Gibbs EM.*Nitric oxide stimulates skeletal muscle glucose transport through a calcium/contraction and phosphatidylinositol-3-kinase-independent pathway.* Diabetes. 1997; 46: 1915–19.

Farrell CL, Pardridge WM. *Blood-brain barrier glucose transporter is asymmetrically distributed on brain capillary endothelial lumenal and ablumenal membranes: an electron microscopic immunogold study.* Proc Natl Acad Sci USA. 1991; 88: 5779–83.

Fernandes R, Suzuki K, Kumagai AK. *Inner blood-retinal barrier GLUT1 in long-term diabetic rats: an immunogold electron microscopic study.* Invest Ophthalmol Vis Sci. 2003; 44: 3150–4.

Fischer Y, Thomas J, Rosen P, Kammermeier H. *Action of metformin on glucose transport and glucose transporter GLUT1 and GLUT4 in heart muscle cells from healthy and diabetic rats.* Endocrinology 1995; 136: 412–20.

Fujiwara R, Nakai T. *Effects of glucose, insulin, and insulin-like growth factor-1 on glucose transport activity in cultured rat vascular smooth muscle cells.* Atherosclerosis. 1996; 127: 49–57.

Galle J, Hansen-Hagge T, Wanner C, Seibold S. *Impact of oxidized low density lipoprotein on vascular cells.* Atherosclerosis. 2006; 185: 219–26.

Gaposchkin CG, Garcia-Diaz JF. *Modulation of cultured brain, adrenal and aortic endothelial cell glucose transport.* Biochim Biophys Acta. 1996; 1285: 255–66.

Gaudreault N, Scriven DR, Moore ED. *Characterization of glucose transporters in the intact coronary artery endothelium in rats: GLUT-2 upregulated by long-term hyperglycaemia.* Diabetologia 2004; 47: 2081–92.

Gaudreault N, Scriven DRL, Moore EDW. *Assymmetric subcellular distribution of glucose transporters in the endothelium of small contractile arteries.* Endothelium. 2006; 13: 317–24.

Gaudreault N, Scriven DRL, Laher I, Moore EDW. *Subcellular characterization of glucose uptake in coronary endothelial cells.* Microvasc Res. 2008; 75: 73–82.

Gerhart DZ, LeVasseur RJ, Broderius MA, Drewes LR. *Glucose transporter localization in brain using light and electron immunocytochemistry.* J Neurosci Res. 1989; 22: 464–72.

Gerritsen ME, Burke TM. *Insulin binding and effects of insulin on glucose uptake and metabolism in cultured rabbit coronary microvessel endothelium.* Proc Soc Exp Biol Med. 1985; 180: 17–23.

Gerritsen ME, Burke TM, Allen LA. *Glucose starvation is required for insulin stimulation of glucose uptake and metabolism in cultured microvascular endothelial cells.* Microvasc Res. 1988; 35: 153–66.

Giardino I, Edelstein D, Brownlee M. *Nonenzymatic glycosylation in vitro and in bovine endothelial cells alters basic fibroblast growth factor activity. A model for intracellular glycosylation in diabetes.* J Clin Invest. 1994; 94: 110–7.

Gjedde A, Crone C. *Blood-brain glucose transfer: repression in chronic hyperglycemia.* Science. 1981; 214: 456–67.

Gonzalez-Flecha B. *Oxidant mechanisms in response to ambient air particles.* Mol Aspects Med. 2004; 25: 169–82.

Gorovits N, Charron MJ. *What we know about facilitative glucose transporters. Lessons from cultured cells, animal models, and human studies.* Biochem Mol Biol Educ. 2003; 31: 163–72.

Gosmanov AR, Stentz FB, Kitabchi AE. *De novo emergence of insulin-stimulated glucose uptake in human aortic endothelial cells incubated with high glucose.* Am J Physiol Endocrinol Metab. 2006; 290: E516–22.

Gregg EW, Yaffe K, Cauley JA, Rolka DB, Blackwell TL, Narayan KM, Cummings SR. *Is diabetes associated with cognitive impairment and cognitive decline among older women?* Arch Intern Med. 2000; 160: 174–80.

Grundy SM, Benjamin IJ, Burke GL, Chait A, Eckel RH, Howard BV, Mitch W, Smith Jr SC, Sowers JR. *Diabetes and cardiovascular disease: a statement for healthcare professionals from the American Heart Association.* Circulation. 1999; 100: 1134–46.

Hadi HA, Suwaidi JA. *Endothelial dysfunction in diabetes mellitus.* Vasc Health Risk Manag. 2007; 3: 853–76.

Halliwell B. *Free-radicals and vascular-disease: how much do we know?* Br Med J. 1993; 307: 885–6.

Hardin CD, Paul RJ. *Metabolism and energetics of vascular smooth muscle.* In: Sperelakis N (ed.) Physiology and Pathophysiology of the Heart. Kluwer Academic, Dordrecht. 1995, 1069–86.

Hardin CD, Roberts TM. *Differential regulation of glucose and glycogen metabolism in vascular smooth muscle by exogenous substrates.* J Mol Cell Cardiol. 1997; 29: 1207–16.

Harik SI, La Manna JC. *Vascular perfusion and blood-brain glucose transport in acute and chronic hyperglycemia.* J Neurochem. 1988; 51: 1924–9.

Hauguel-de-Mouzon S, Challier JC, Kacemi A, Cauzac M, Malek A, Girad J. *The GLUT3 glucose transporter isoform is differentially expressed within human placental cell types.* J Clin Endocrinol Metab. 1997; 82: 2689–94.

Hertz MM, Paulson OB, Barry DI, Christiansen JS, Svendsen PA. *Insulin increases glucose transfer across the blood-brain-barrier in man.* J Clin Invest. 1981; 67: 597–604.

Hingorani V, Brecher P. *Glucose and fatty acid metabolism in normal and diabetic rabbit cerebral microvessels.* Am J Physiol Endocrinol Metab. 1987; 252: E648–53.

Howard RL. *Down-regulation of glucose transport by elevated extracellular glucose concentrations in cultured rat aortic smooth muscle cells does not normalize intracellular glucose concentrations.* J Lab Clin Med. 1996; 127: 504–15.

Jacob RJ, Fan X, Evans ML, Dziura J, Sherwin RS. *Brain glucose levels are elevated in chronically hyperglycemic diabetic rats: no evidence for protective adaptation by the blood brain barrier.* Metabolism. 2002; 51: 1522–4.

James DE, Strube M, Mueckler M. *Molecular cloning and characterization of an insulin-regulatable glucose transporter.* Nature. 1989; 338: 83–7.

Kahn AM, Lichtenberg RA, Allen JC, Seidel CL, Song T. *Insulin stimulated glucose transport inhibits Ca2+ influx and contraction in vascular smooth muscle.* Circulation. 1995; 92: 1597–1603.

Kaiser N, Sasson S, Feener EP, Boukobzavardi N, Higashi S, Moller DE, Davidheiser S, Przybylski RJ, King GL. *Differential regulation of glucose transport and transporters by glucose in vascular endothelial and smooth muscle cells.* Diabetes. 1993; 42: 80–9.

Kandror KV, Pilch PF. *Compartmentalization of protein traffic in insulin-sensitive cells.* Am J Physiol. 1996; 271: E1–14

Kapitulnik J. *Bilirubin: an endogenous product of heme degradation with both cytotoxic and cytoprotective properties.* Mol Pharmacol. 2004; 66: 7737–79.

Kihara S, Ouchi N, Funahashi T, Shinohara E, Tamura R, Yamashita S, Matuzawa Y. *Troglitazone enhances glucose uptake and inhibits mitogen-activated protein kinase in human aortic smooth muscle cells.* Atherosclerosis. 1998; 136: 163–8.

Klepper J, Wang D, Fischbarg J, Vera JC, Jartour IT, O'Driscoll KR, Devivo DC. *Defective glucose transport across blood brain tissue barriers: a newly recognized neurological syndrome.* Neurochem Res. 1999; 24: 587–94.

Knott RM, Robertson M, Muckersie E, Forrester JV. *Regulation of glucose transporters (GLUT-1 and GLUT-3) in human retinal endothelial cells.* Biochem J. 1996; 318: 313–7.

Kreutzfeldt A, Spahr R, Mertens S, Siegmund B, Piper HM. *Metabolism of exogenous substrates by coronary endothelial cells in culture.* J Mol Cell Cardiol. 1990; 22: 1393–1404.

Kumagai AK. *Glucose transport in brain and retina: implications in the management and complications of diabetes.* Diabetes Metab Res Rev. 1999; 15: 261–73.

Kumagai AK, Glasgow BJ, Pardridge WM.*Glut1 glucose transporter expression in the diabetic and nondiabetic human eye.* Invest Ophthamol Vis Sci. 1994; 35: 2887–94.

Kumagai AK, Kang YS, Boado RJ, Pardridge WM. *Upregulation of blood-brain barrier GLUT1 glucose transporter protein and mRNA in experimental chronic hypoglycemia.* Diabetes. 1995; 44: 1399–1404.

Lee IK, Kim HS, Bae JH. *Endothelial dysfunction: its relationship with acute hyperglycaemia and hyperlipidemia.* Int J Clin Pract. 2002; Suppl.129: 59–64.

Loike JD, Cao L, Brett J, Ogawa S, Silverstein SC, Stern D. *Hypoxia induces glucose transporter expression in endothelial cells.* Am J Physiol Cell Physiol. 1992; 263: C326–33.

Lorenzi M, Cagliero E, Toledo S. *Glucose toxicity for human endothelial cells in culture. Delayed replication, disturbed cell cycle, and accelerated death.* Diabetes. 1985; 34: 621–7.

Machado UF, Schaan BD, Seraphim PM. *Transportadores de glicose na síndrome metabólica.* Arq Bras Endocrinol Metab. 2006; 50: 177–89.

MacKenzie CJ, Wakefield JM, Cairns F, Dominiczak AF, Gould GW. *Regulation of glucose transport in aortic smooth muscle cells by cAMP and cGMP.* Biochem J. 2001; 353: 513–9.

Mandarino LJ, Finlayson J, Hassell JR. *High glucose downregulates glucose transport activity in retinal capillary pericytes but not endothelial cells.* Invest Ophthamol Vis Sci. 1994; 35: 964–72.

Mann GE, Yudilevich DL, Sobrevia L. *Regulation of amino acid and glucose transporters in endothelial and smoothe muscle cells.* Physiol Rev. 2003; 83: 183–252.

Mantych GJ, James DE, Devaskar SU. *Jejunal kidney glucose transporter isoform (GLUT-5) is expressed in the human bloodbrain- barrier.* Endocrinology. 1993a; 132: 35–40.

Mantych GJ, Hageman GS, Devastu SU. *Characterization of glucose transporter isoforms in the adult and developing human eye.* Endocrinology. 1993b; 133: 600–7.

Marcus RG, England R, Nguyen K, Charron MJ, Briggs JP, Brosius FC3. *Altered renal expression of the insulin-responsive glucose transporter GLUT4 in experimental diabetes mellitus.* Am J Physiol. 1994; 267: 816–24.

Matthaei S, Hamann A, Klein HH, Benecke H, Krey-Mann G, Flier JS, Greten H. *Association of metformin's effect to increase insulin-stimulated glucose transport with potentiation of insulininduced translocation of glucose transporters from intracellular pool to plasma membrane in rat adipocytes.* Diabetes. 1991; 40: 850–7.

McCall AL, Millington WR, Wurtman RJ. *Metabolic fuel and amino acid transport into the brain in experimental diabetes mellitus.* Proc Natl Acad Sci USA. 1982; 79: 5406–10.

McCall AL, Gould JB, Ruderman NB. *Diabetes-induced alterations of glucose metabolism in rat cerebral microvessels.* Am J Physiol. 1984; 247: E462–7.

McCall AL, Fixman LB, Fleming N, Tornheim K, Chick W, Ruderman NB. *Chronic hypoglycemia increases brain glucose transport.* Am. J. Physiol. 1986; 251: E442–7.

McCall AL, Van Bueren AM, Huang L, Stenbit A, Celnik E, Charron MJ. *Forebrain endothelium expresses GLUT4, the insulin responsive glucose transporter.* Brain Res. 1997; 744: 318–26.

Mertens S, Noll T, Spahr R, Kruetzfeldt A, Piper HM. *Energetic response of coronary endothelial cells to hypoxia.* Am J Physiol Heart Circ Physiol. 1990; 258: H689–94.

Mooradian AD, Morin AM, Cipp LJ, Haspel HC. *Glucose transport is reduced in the blood-brain barrier of aged rats.* Brain Res. 1991; 551: 145–59.

Namba H, Lucignani G, Nehlig A, Patlak C, Pettigrew K, Kennedy C, Sokoloff L. *Effects of insulin on hexose transport across blood-brain-barrier in normoglycemia.* Am J Physiol Endocrinol Metab. 1987; 252: E299–303.

Nishikawa T, Edelstein D, Du XL, Yamagishi S, Matsumura T, Kaneda Y, Yorek MA, Beebe D, Oates PJ, Hammes HP, Giardino I, Brownlee M. *Normalizing mitochondrial superoxide production blocks three pathways of hyperglycaemic damage.* Nature. 2000; 404: 787–90.

Nitenberg A, Cosson E, Pham I. *Postprandial endothelial dysfunction: role of glucose, lipids and insulin.* Diabetes Metab. 2006; 32 (Spec No. 2): S28–33.

O'Brien RM, Granner DK. *Regulation of gene expression by insulin.* Physiol Rev. 1996; 76: 1109–61.

Pan M, Wasa M, Souba WW. *Tumor necrosis factor stimulates system x^-_{AG}transport activity in human endothelium.* J Surg Res. 1995; 58: 659–64.

Pardridge WM, Boado RJ, Farrell CR. *Brain-type glucose transporter (GLUT-1) is selectively localized to the blood-brain barrier. Studies with quantitative western blotting and in situ hybridization.* J Biol Chem. 1990a; 256: 18035–40.

Pardridge WM, Triguero D, Farrell CR. *Downregulation of blood-brain barrier glucose transporter in experimental diabetes.* Diabetes. 1990b; 39: 1040–4.

Park JL, Loberg RD, Duquaine D, Zhang H, Deo BK, Ardanaz N, et al. *GLUT4 facilitative glucose transporter specifically and differentially contributes to agonist induced vascular reactivity in mouse aorta.* Arterioscler Thromb Vasc Biol. 2005; 25: 1596–602.

Parke DV, Sapota A. *Chemical toxicity and reactive oxygen species.* Int J Occup Med Environ Health. 1996; 9: 331–40.

Parra M, Yudilevich DL, Mann GE, Pedley KC, Nicolaides K, Pearson JD, Sobrevia L. *Modulation of glucose transport in human fetal vein endothelial cells.* J Physiol. 1998; 506: 34–5.

Paulson OB, Hasselbalch SG. *Blood-brain barrier transport of glucose: adaptation to changes in blood glucose levels.* Nutr Metab Cardiovasc Dis. 1997; 7: 217–24.

Pekala P, Marlow M, Heuvelman D, Connoly D. *Regulation of hexose transport in aortic endothelial cells by vascular permeability factor and tumor necrosis factor-α, but not by insulin.* J Biol Chem. 1990; 265: 18051–4.

Pelligrino DA, LaManna JC, Duckrow RB, Bryan RM, Harik SI. *Hyperglycemia and blood-brain barrier glucose transport.* J Cereb Blood Flow Metab. 1992; 12: 887–99.

Pessler D, Rudich A, Bashan N. *Oxidative stress impairs nuclear proteins binding to the insulin responsive element in the GLUT4 promoter.* Diabetologia. 2001; 44: 2156–64.

Pouliot JF, Beliveau R. *Palmitoylation of the glucose transporter in blood-brain barrier capillaries.* Biochim Biophys Acta. 1995; 1234: 191–6.

Quinn LA, McCumbee WD. *Regulation of glucose transport by angiotensin II and glucose in cultured vascular smooth muscle cells.* J Cell Physiol. 1998; 177: 94–102.

Rebolledo OR, Actis Dato SM. *Postprandial hyperglycemia and hyperlipidemia-generated glycoxidative stress: its contribution to the pathogenesis of diabetes complications.* Eur Rev Med Pharmacol Sci. 2005; 9: 191–208.

Regina A, Roux F, Revest PA. *Glucose transport in immortalized rat brain capillary endothelial cells in vitro: transport activity and GLUT1 expression.* Biochim Biophys Acta. 1997; 1335: 135–43.

Sasson S, Gorowits N, Joost HG, King GL, Cerasi E, Kaiser N. *Regulation by metformin of the hexose transport system in vascular endothelial and smooth muscle cells.* Br J Pharmacol. 1996; 117: 1318–24.

Schalkwijk CG, Stehouwer CD. *Vascular complications in diabetes mellitus: the role of endothelial dysfunction.* Clin Sci (Lond). 2005; 109: 143–59.

Shepherd PR, Kahn BB. *Glucose transporters and insulin action–implications for insulin resistance and diabetes mellitus.* N Engl J Med. 1999; 341: 248–57.

Slot JW, Moxley R, Geuze HJ, James DE. *No evidence for expression of the insulin-regulatable glucose transporter in endothelial cells.* Nature. 1990; 346: 369–71.

Sobrevia L, Mann GE. *Dysfunction of the endothelial nitric oxide signalling pathway in diabetes and hyperglycaemia.* Exp Physiol. 1997; 82: 423–52.

Sone H, Deo BK, Kumagai AK. *Enhancement of glucose transport by vascular endothelial growth factor in retinal endothelial cells.* Invest Ophthalmol Vis Sci. 2000; 41: 1876–84.

Standley PR, Rose KA. *Insulin and insulin-like growth factor-1 modulation of glucose transport in arterial smooth muscle cells: implication of GLUT-4 in the vasculature.* Am J Hypertens. 1994; 7: 357–62.

Stenina OI. *Regulation of vascular genes by glucose.* Curr Pharm Des. 2005; 11: 2367–81.

Stratton IM, Adler AI, Neil HA, Matthews DR, Manley SE, Cull CA, Hadden D, Turner RC, Holman RR. *Association of glycaemia with macrovascular and microvascular complications of type 2 diabetes (UKPDS 35): prospective observational study.* Br Med J. 2000; 321: 405–12.

Tada H, Thompson CI, Recchia FA, Loke KE, Ochoa M, Smith CJ, Shelsely EG, Kaley G, Hintze TH. *Myocardial glucose uptake is regulated via endothelial nitric oxide synthase in Langendorff mouse heart.* Circ Res. 2000; 86: 270–4.

Takagi H, King GL, Aiello LP. *Hypoxia upregulates glucose transport activity through an adenosine-mediated increase in GLUT1 expression in retinal capillary endothelial cells.* Diabetes. 1998; 47: 1480–8.

Takakura Y, Kuentzel SL, Raub TJ, Davies A, Baldwin SA, Borchardt RT. *Hexose uptake in primary cultures of bovine brain microvessel endothelial cells. I. Basic characteristics and effects of D-glucose and insulin.* Biochim Biophys Acta. 1991; 1070: 1–10.

Takata K, Hirano H, Kasahara M. *Transport of glucose across the blood-tissue barriers.* Int Rev Cytol. 1997; 172: 1–53.

Tao F, Gonzalez-Flecha B, Kobzik L. *Reactive oxygen species in pulmonary inflammation by ambient particulates.* Free Radic Biol Med. 2003; 35: 327–40.

Taylor R, Agius L. *The biochemistry of diabetes.* Biochem J. 1988; 250: 625–50.

Thomas J, Linssen M, Van Der Vusse GJ, Hirsch B, Rosen P, Kammermeier H, Fischer Y. *Acute stimulation of glucose transport by histamine in cardiac microvascular endothelial cells.* Biochim Biophys Acta. 1995; 1268: 88–96.

Thornalley PJ, McLellan AC, Lo TW, Benn J, Sonksen PH. *Negative association between erythrocyte reduced glutathione concentration and diabetic complications.* Clin Sci (Lond). 1996; 91: 575–82.

Totary-Jain H, Naveh-Many T, Riahi Y, Kaiser N, Eckel J, Sasson S. *Calreticulin destabilizes glucose transporter-1 mRNA in vascular endothelial and smooth muscle cells under high-glucose conditions.* Circ Res. 2005; 97: 1001–8.

Tsuneki H, Sekizaki N, Suzuki T, Kobayashi S, Wada T, Okamoto T, Kimura I, Sasaoka T. *Coenzyme Q10 prevents high glucose-induced oxidative stress in human umbilical vein endothelial cells.* Eur J Pharmacol. 2007; 566: 1–10.

Vilaro S, Palacín M, Pilch PF, Testar X, Zorzano A. *Expression of an insulin-regulatable glucose carrier in muscle and fat endothelial cells.* Nature. 1989; 342: 798–800.

Viñals F, Gross A, Testar X, Palacín M, Rosen P, Zorzano A. *High glucose concentrations inhibit glucose phosphorylation, but not glucose transport, in human endothelial cells.* Biochim Biophys Acta. 1999; 1450: 119–29.

Vincent AM, Russell JW, Low P, Feldman EL. *Oxidative stress in the pathogenesis of diabetic neuropathy.* Endocr Rev. 2004; 25: 612–28.

Wei M, Gaskill SP, Haffner SM, Stern MP. *Effects of diabetes and level of glycemia on all-cause and cardiovascular mortality.* The San Antonio Heart Study. Diabetes Care. 1998; 21: 1167–72.

Wiernsperger N, Nivoit P, De Aguiar LG, Bouskela E. *Microcirculation and the metabolic syndrome.* Microcirculation. 2007; 14: 403–38.

Wood IS, Trayhurn P. *Glucose transporters (GLUT and SGLT1): expanded families of sugar transport proteins.* Br J Nutr. 2003; 89: 3–9.

Wright EM, Hirayama BA, Loo DF. *Active sugar transport in health and disease.* J Intern Med. 2007; 261: 32–43.

Yamagishi SI, Nakamura K, Matsui T, Ueda SI, Imaizumi T. *Role of postprandial hyperglycaemia in cardiovascular disease in diabetes.* Int J Clin Pract. 2007; 61: 83–7.

Yu Y, Lyons TJ. *A lethal tetrad in diabetes: hyperglycemia, dyslipidemia, oxidative stress, and endothelial dysfunction.* Am J Med Sci. 2005; 330: 227–32.

Chapter 8
Natural Polyphenols as Anti-Oxidant, Anti-Inflammatory and Anti-Angiogenic Agents in the Metabolic Syndrome

Rita Negrão and Ana Faria

Contents

Abstract The world is increasingly threatened by chronic diseases, related to lifestyle, stress, lack of physical exercise and excessive intake of high caloric diets. Almost all chronic diseases have in common an increased level of oxidative stress and inflammatory state.

R. Negrão (✉)
Biochemistry Department (U38-FCT), Faculty of Medicine, University of Porto, Al. Prof. Hernâni Monteiro, 4200-319 Porto, Portugal
e-mail: ritabsn@med.up.pt

R. Soares, C. Costa (eds.), *Oxidative Stress, Inflammation and Angiogenesis in the Metabolic Syndrome*, DOI 10.1007/978-1-4020-9701-0_8,

A wide variety of dietary plants including vegetables, fruits and beverages, like tea and wine, are rich in polyphenols. These compounds have attracted much attention with regard to human health as they seem to protect the establishment of several diseases with high prevalence in western countries, like cancer, cardiovascular disorders and diabetes. Here we will discuss general aspects of polyphenols classification, metabolism and bioavailability, their role as anti-oxidants and modulators of cell signalling, focusing especially in inflammation and angiogenic pathways. Recent reports suggesting beneficial effects of these compounds on metabolic syndrome will also be referred.

Keywords Anti-angiogenic · Anti-inflammatory · Anti-oxidant · Bioavailability · Flavonoid · Metabolic syndrome · Polyphenols

8.1 Introduction

Polyphenolic compounds are ubiquitously found in the plant kingdom. Products of plant secondary metabolism, they are responsible for plant survival face to environmental threats. Polyphenols provide color to leaves, flowers and fruit, possess anti-microbial and anti-fungal properties, exert insect feeding deterrence, screening from solar UV radiation damage, chelation of toxic heavy metals and anti-oxidant protection from free radicals generated during the photosynthetic process (Cooper-Driver and Bhattacharya 1998). Apart from their physiological roles in plants, flavonoids are important components in the human diet, although they are considered as non-nutrients. Several epidemiological studies provide support for the association between consumption of fruit, vegetables and certain beverages rich in polyphenols, such as tea and red wine, and health promoting effects (Kaur and Kapoor 2001; Van Duyn and Pivonka 2000). Numerous studies have indicated that individuals with a high consumption of fruits and vegetables have a reduced incidence of age-associated illness such as neuro-degenerative diseases (Aquilano et al. 2008; Lau et al. 2006; Singh et al. 2008). Recently, correlation between reduced incidence of death from cardiovascular and coronary heart disease and a regular intake of a diet rich in flavonoids has been demonstrated (Corder et al. 2006; Mink et al. 2007). Also, flavonoid consumption has been correlated with the reduction in cancer incidence, namely renal cell carcinoma (Bosetti et al. 2007), colorectal (Rossi et al. 2006), breast (Bosetti et al. 2005) and ovarian cancer (Rossi et al. 2008).

In addition, flavonoids exhibit a wide range of biological activities, including anti-inflammatory, anti-viral, anti-bacterial, anti-ulcer, anti-osteoporotic and anti-allergic actions (Cushnie and Lamb 2005; Di Carlo et al. 1999). The beneficial effects of polyphenols are mainly attributed to their capacity to counteract conditions of oxidative stress that accompany these pathologies. Several polyphenols have demonstrated to have clear anti-oxidant properties *in vitro* as they can act as chain breakers or radical scavengers depending on their chemical structures (Hu et al. 1995; Rice-Evans 2001). Nevertheless, restricting the concept of flavonoid biological effects to their anti-oxidant properties appears to be a simplistic way to conceive their activity.

Currently, over 8000 polyphenolic compounds that share the presence of an aromatic ring with two or more hydroxyl groups linked directly to the aromatic ring are known. Mono-phenols (e.g. *p*-coumaric acid) are not polyphenols, but they share many of polyphenol properties and characteristics and are therefore considered as "functional polyphenols". Polyphenols can be broadly divided into two categories: flavonoid and non-flavonoid polyphenols, flavonoids being definitely the most studied category (Table 8.1).

8.2 Natural Occurrence and Intake

Polyphenols are widespread, virtually present in all foods and beverages of plant origin, such as fruits, vegetables, tea, cocoa and wine. For several reasons, it is extremely difficult to estimate the daily average intake of polyphenols, such as the extensive diversity of chemical structures that makes the estimation of polyphenol content in foods complex, the analytical method used for these estimations, variation of content with geographical region, harvesting and season, as well as dietary habits of people. Few estimations of dietary intake are available and the attained values are somehow different. A recent study compared the dietary fiber intake of two European populations: one from Denmark and other from Spain. The total dietary fiber intake was 31% higher in Murcia (Spain) than in Copenhagen (Denmark)(Tabernero et al. 2007). Another study carried out with the Fijian population showed that this population generally had low intakes of total phenols (275 mg/day), and total flavonoids (17.5 mg/day), but high intake of total carotenoids (20 mg/day), in comparison with the intakes of other populations reported in literature (Lako et al. 2006). Ovaskainen and co-workers (Ovaskainen et al. 2008) have conducted a study with Finnish adults and the total intake of polyphenols was around 850 mg/day. In this population, coffee and cereals were the main contributors to total polyphenol intake. Scalbert and Williamson have already proposed that the total dietary intake is about 1 g/day, which is much higher than that of all other known dietary anti-oxidants, about 10 times higher than that of vitamin C and 100 times higher than those of vitamin E and carotenoids (Scalbert and Williamson 2000). It is estimated that one-third of the total polyphenols ingested are phenolic acids and other non-flavonoids and the further two thirds are flavonoids.

8.3 Flavonoid Chemical Structure

The basic flavonoid structure is the flavan nucleus, which consists of a C6-C3-C6 structure constituted by two aromatic rings linked by an heterocyclic ring, labelled A, B and C (Fig. 8.1). The various classes of flavonoids differ in the level of oxidation and pattern of substitution at the C ring, while individual compounds within a class differ in the arrangements of hydroxyl, methoxyl and glycosidic

Table 8.1 Chemical structures of main groups of polyphenols

Groups of polyphenols			Molecular structure	Functional groups and examples
Polyphenols	Flavonoid	Flavonols	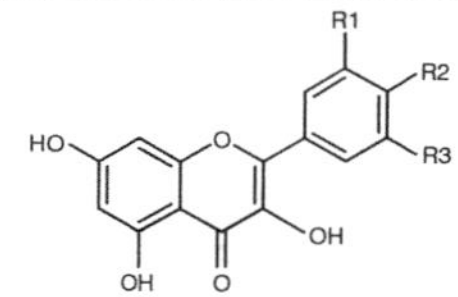	R1=H; R2=OH; R3=H: kaempferol R1=OH; R2=OH; R3=H: quercetin R1=OH; R2=OH; R3=OH: myricetin
		Flavan-3-ols	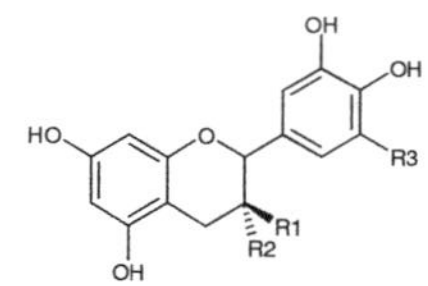 Monomeric	R1=H; R2=OH; R3=H: (+)-catechin R1=OH; R2=H; R3=H: (–)-epicatechin R1=OH; R2=OH; R3=OH: (+)-gallocatechin
			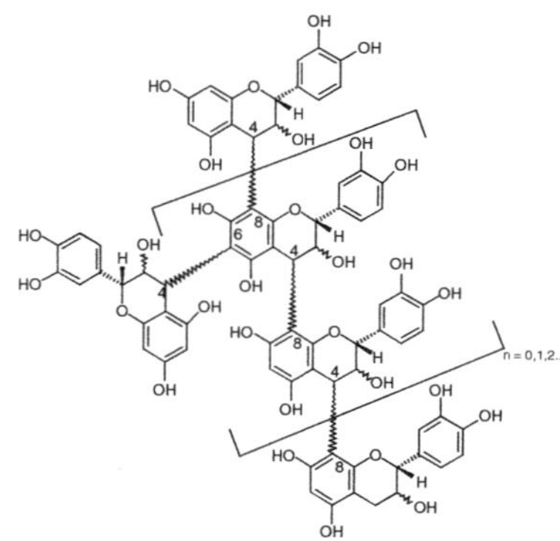 Proanthocyanidins	
		Flavones	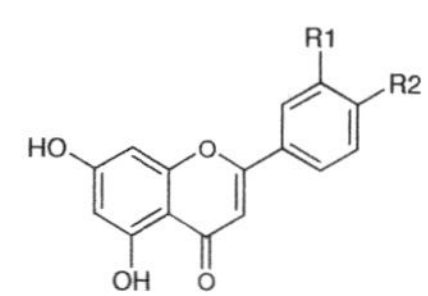	R1=H; R2=OH: apigenin R1=OH; R2=OH: luteolin

Table 8.1 (continued)

Groups of polyphenols		Molecular structure	Functional groups and examples
	Flavanones	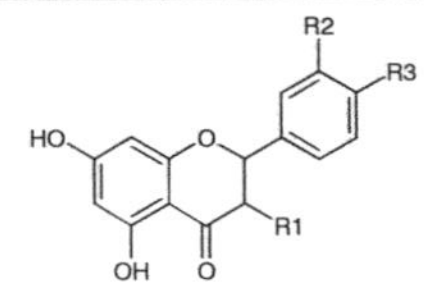	R1=H; R2=H; R3=OH: naringenin R1=OH; R2=OH; R3=OMe: hesperetin
	Isoflavones	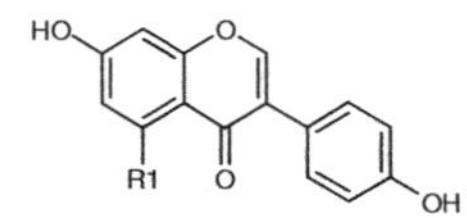	R1=H: daidzein R1=OH: genistein
	Anthocyanidins	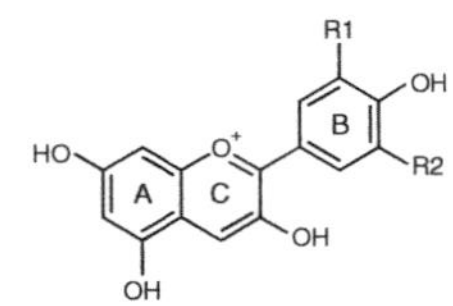	R1=H; R2=H: pelargonidin R1=OH; R2=H: cyanidin R1=OH; R2=OH: delphinidin R1=OMe; R2=OH: petunidin R1=OMe; R2=OMe: malvidin
	Chalcones	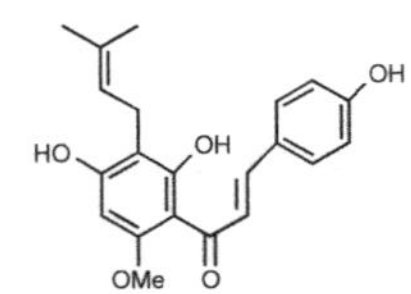	xanthohumol

Table 8.1 (continued)

Groups of polyphenols			Molecular structure	Functional groups and examples
	Non-flavonoid	Phenolic acids	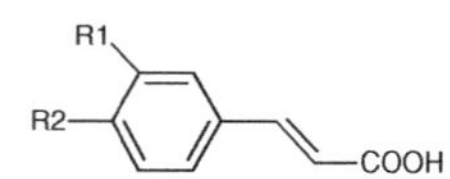 Hydroxycinnamic acids	R1=H; R2=OH: *p*-coumaric acid R1=OH; R2=OH: caffeic acid R1=OMe; R2=OH: ferulic acid
			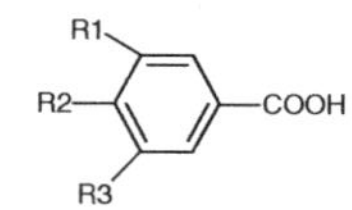 Hydroxybenzoic acids	R1=OH; R2=OH; R3=H: protocatechuic acid R1=OH; R2=OH; R3=OH: gallic acid
		Stilbenes	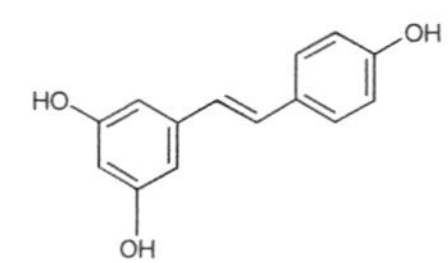	resveratrol
		Lignans	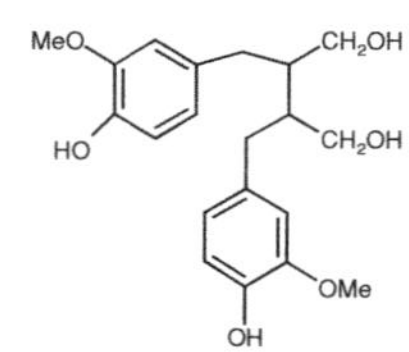	secoisolariciresinol

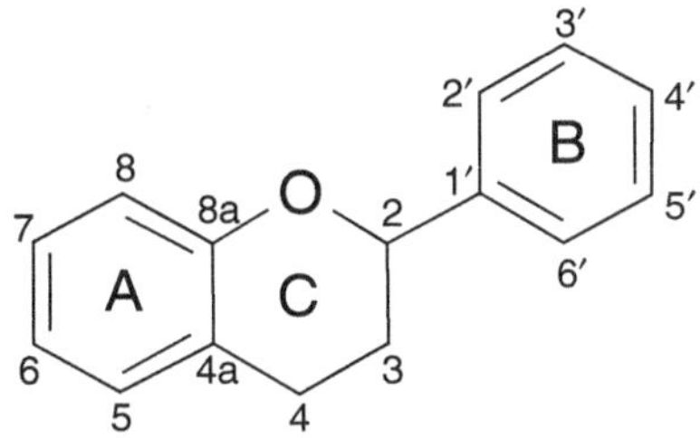

Fig. 8.1 Basic structure of flavan nucleus

side groups (Table 8.1). Dietary flavonoids exist primarily as 3-*O*-glycosides and polymers (Hammerstone et al. 2000).

Flavonoids are known to be powerful anti-oxidants. Their mechanisms of action can be direct or indirect: scavenging reactive free radicals and chelating metal ions are the most significant direct effects; reduction of free radical production through inhibition of enzymes and regeneration of membrane-bound anti-oxidants such as α-tocoferol are examples of indirect mechanisms.

In 1990, Bors et al. (Bors et al. 1990) described for the first time the three structural features that determine the radical scavenging and/or anti-oxidative potential of flavonoids: (a) the *o*-dihydroxyl (catechol) structure in the B ring; (b) the C2-C3 double bond in conjugation with a 4-oxo group; (c) the additional presence of both 3- and 5- hydroxyl groups. Over the years, structure-activity relationship research has generated several consistent lines of evidence supporting the role of these specific structural features as requisites for radical scavenging, chelation and oxidant activity (Amic et al. 2007; Heim et al. 2002).

8.3.1 Structural Features for Anti-Oxidant Activity

The configuration and total number of hydroxyl groups substantially influence several mechanisms of anti-oxidant activity. Free radical scavenging capacity is primarily attributed to the high reactivities of hydroxyl substituents. Hydroxyl groups on the B-ring donate hydrogen and an electron to hydroxyl, peroxyl and peroxynitrite radicals, stabilizing them and giving rise to a relatively stable flavonoid radical (*o*-semiquinone radical) through facilitating electron delocalization (Arora et al. 1998). Among homologous structures of flavones, peroxyl and hydroxyl scavenging increases linearly according to the number of hydroxyl groups (Cao et al. 1997). Flavones lacking these features form relatively unstable radicals and are weak scavengers. Despite the disparity among methods of assessing activity, there is a broad agreement that favourable position of hydroxyl groups, like the catechol moiety in the B-ring, could be a prerequisite for the stability of phenoxyl radical.

Polymethylated flavonoids present different anti-oxidant activities than polyhydroxylated most likely due to differences in hydrophobicity and molecular planarity. *O*-methylated and *O*-glycosylated quercetin derivatives are less potent than quercetin itself (Dugas et al. 2000). The lipophilicity and membrane partitioning ability afforded by methoxy groups can be physiologically relevant parameters for anti-oxidant activity (Ollila et al. 2002; van Acker et al. 1996). *O*-methylation improves anti-oxidant activity in some microsomal systems (Walle 2007). However, in these systems, flavonoids may be acting by several mechanisms (recycling endogenous microsomal α-tocoferol, chelating metals and undergoing biotransformation by cytochrome P450) which can explain this greater activity.

One of the characteristic features of some classes of flavonoids is the presence of a 2-3 double bond in conjugation with a 4-oxo group. Despite the importance it seems to have, the net result on anti-oxidant activity is not so clear. Results obtained with flavonols and anthocyanidins suggest that these may not be essential provided that other structural criteria are fulfilled (Rice-Evans et al. 1996). The flavanone taxofolin has a weaker TEAC value in respect to the flavonol quercetin, due to the lack of 2,3-double bond with 4-oxo conjugation (Rice-Evans et al. 1996). The majority of the reports suggest that flavonoids lacking one or both of these features are less potent anti-oxidants than those with both elements.

There is not a straight correlation between the presence of a carbohydrate moiety and the anti-oxidant activity. Frequently aglycones are more potent anti-oxidant than their corresponding glycosides (Ratty and Das 1988; Zhou et al. 2005). Not only the presence and total number of sugar, but also the position and structure of the sugar is of considerable importance. In the diet, glycosidic moieties are usually in 3- or 7- positions. Although glycosides are usually weaker anti-oxidants, their bioavailability is often enhanced due to the glucose moiety (Hollman et al. 1999). Despite the sugar moiety occupying one free hydroxyl group that could be used for hydrogen abstraction and radical scavenging, its presence is also capable of diminishing the coplanarity of B-ring relative to the rest of the flavonoid and introducing hydrophilicity, altering access to lipid peroxyl and alkoxyl radicals. Nevertheless, the importance of sugar moiety on flavonoid anti-oxidant activity is questionable since some reports claim that flavonoids are deglycosylated at gut level (Gee et al. 2000; Wilkinson et al. 2003).

Most of the flavonoids ingested in western cultures are polymerized flavonoids. Structure-activity relationships of these molecules are poorly understood. Procyanidin dimers and trimers are more effective against superoxide than monomeric flavonoids, but differ little between them (Vennat et al. 1994). Tetramers demonstrated greater activity and hexamers and heptamers had even superior activities (Vennat et al. 1994). It appears that anti-oxidant activity of procyanidins increases with the degree of polymerization. The extensive conjugation between 3-OH and B-ring catechol groups is responsible for great scavenging properties by increasing radical stability (Castillo et al. 2000). However, a ranking of structure-activity relationship for procyanidins has not been established.

8.3.2 Pro-Oxidant Activity

While the ability of flavonoids to act as anti-oxidants has been undoubtedly demonstrated, some focus has been given to the pro-oxidant activity of these compounds *in vitro* (Cotelle 2001; Galati and O'Brien 2004; Heim et al. 2002). Pro-oxidant activity is though to be directly proportional to the total number of hydroxyl groups (Cao et al. 1997). A series of mono- and dihydroxyflavonoids did not present detectable pro-oxidant activity, while multiple hydroxyl groups, especially in the B-ring, significantly increased production of hydroxyl radicals in a Fenton system (Hanasaki et al. 1994). The presence of three hydroxyl groups (pyrogallol structure) in the A-ring has also been reported to promote hydrogen peroxide production (Hodnick et al. 1986). As well, the effect of 18 flavonoids and related flavonoid compounds was studied on the DNA damage induced by nitric oxide (NO), peroxynitrite and nitroxyl anion. Most of the tested flavonoids inhibited DNA strand breakage. Only flavonoids having an *o*-trihydroxyl group, either in the B-ring or in the A-ring, acted as pro-oxidants (Ohshima et al. 1998). There is also evidence that the unsaturated 2,3-bond and the 4-oxo arrangement of flavones may promote the formation of ROS induced by divalent copper, in the presence of oxygen (Cao et al. 1997). Bors et al. have suggested that the stability of flavonoid phenoxyl radical is sometimes questionable and may give rise to pro-oxidant action (Bors et al. 1995). The structural advantages to radical stability that increase anti-oxidant action, such as 3,4-catechol, 3-OH and conjugation between A- and B- rings may modulate adverse effects of flavonoids.

Glycosylation and methylation of hydroxyl groups soothe the pro-oxidant behaviour of flavonoids (Cao et al. 1997). Adverse oxidative effects could be moderated *in vivo* by catechol-*O*-methyltransferase (COMT) and other hepatic methyltransferases, but this hypothesis is remote since flavonoids directly inhibit COMT (Lu et al. 2003). In spite of the tendency of flavonoid with multiple hydroxyl groups to promote cellular damage, metabolic alterations of structure may attenuate the reactivity of such compounds *in vivo*.

8.4 Absorption, Bioavailability and Metabolism of Flavonoids

Knowledge about absorption, pharmacokinetics, biotransformation and relative activities of metabolites is a critical determinant to understand the biological effects of flavonoids in organisms. Although flavonoids anti-oxidant activity *in vitro* is well established and supported under different circumstances of oxidative stress, anti-oxidant potential of these compounds *in vivo* is still an unclear issue.

To reach the knowledge about flavonoid biological actions in the human organism it is important to clarify: the extent as well as the local of absorption; the pharmacokinetics; the metabolites produced and their biological actions plus structure activity relationships. It is expected that absorption varies substantially with food matrix, dosage, vehicle of administration, the antecedent diet, interindividual and sex differences, and microbial population. A compound being dispersed within a food matrix has a very high surface area of contact with the aqueous phase. In contrast, an extracted flavonoid compressed into a pill has to be actively dispersed. Another factor limiting availability of flavonoids is their ability to complex with proteins (Gonçalves et al. 2007; Soares 2007); in contrast, ethanol may improve bioavailability (Dragoni et al. 2006) as evidenced by the increased uptake of polyphenols of red wine as compared with levels resulting after ingestion of alcohol-free red wine (Scholz and Williamson 2007). Since most of dietary flavonoids occur in food as *O*-glycosides (Hammerstone et al. 2000), it was speculated that those compounds needed to be deglycosylated in order to be absorbed. There are two β-endoglucosidases in the human small intestine capable of flavonoid glycoside hydrolysis (Day et al. 2000; Leese and Semenza 1973). Several authors support the need to deglycosylate flavonoids for absorption (Gee et al. 2000; Wilkinson et al. 2003). However, it has also been suggested that the location and structure of the sugar moiety may influence absorption (Hollman et al. 1999; Shimoi et al. 1998; Wu et al. 2005). The variations in gene expression and protein abundance of important components that rule absorption, distribution, biotransformation and excretion, such as specific transporters or metabolic enzymes, also appreciably alter bioavailability. Even with low uptake rates, flavonoids that are absorbed may undergo three forms of intracellular metabolism: (a) conjugation with thiols, particularly glutathione; (b) oxidative metabolism; (c) cytochrome P450 metabolism. Metabolic modifications of flavonoids will alter their "classical" anti-oxidant nature. The circulating forms of flavonoids are mainly glucuronides, sulphates, and *O*-methylated metabolites, which are believed to be the most likely to exert bioactivity and express beneficial effects in humans and other animals (Spencer et al. 2001a, b). Some studies have indicated that glucuronides, sulphates, and *O*-methylated forms may participate directly in plasma anti-oxidant reactions by scavenging reactive oxygen and nitrogen species in the circulation in a similar manner to their parent without metabolic modifications (Shirai et al. 2001; Terao et al. 2001), albeit with reduced efficacy. Furthermore, concentration of flavonoids and their metabolite forms *in vivo*, as well as circulating levels, may be too low to be relevant. It remains unclear if the flavonoid metabolites express biological activities at the cellular level, although recent research has focused on this issue (Lambert et al. 2007; Loke et al. 2008; Steffen et al. 2008). The cellular effects of these forms will ultimately depend on the extent to which they associate with cells, either by interactions at the membrane or uptake into the cytosol.

8.5 Inflammation and Polyphenols

The process of acute inflammation is initiated within the microvessels near the injured tissue, which alter exudation of plasma proteins and leukocytes (neutrophils, eosinophils and macrophages) into the surrounding tissue, causing the characteristic swelling, reddened and increased heat associated with inflammation. Once in the tissue, the cells migrate along a chemotactic gradient to reach the site of injury, where they attempt to repair the tissue. Several biochemical inflammatory mediators, act in parallel to propagate and mature the inflammatory response. Most inflammatory modulators have short half-lives, helping to quickly cease the inflammatory response once the stimulus has been removed. There are several inflammatory modulators such as, vasoactive amines (histamine and 5-hydroxytryptamine);

adhesion molecules (intercellular and vascular cell adhesion molecules, ICAM 1, and V-CAM), selectins; lipid-derived eicosanoids (prostaglandin $E_2$and I_2, PGE_2, PGI_2), leukotriene B4 (LTB4), LTC4; cytokines (tumour necrosis factor α, TNFα, interleukin-1β (IL-1β), IL-6, IL-10 and chemokines (IL-8, monocyte-chemoattractant protein-1, MCP-1, macrophage inflammatory molecule 1α, MIP1α) (Bengmark 2004; Santangelo et al. 2007).

The inflammatory response is a complex self-limiting process precisely regulated to prevent excessive damage to the host. When control mechanisms do not function properly, they originate a pathological chronic inflammatory state. Chronic inflamed tissue is characterised by the infiltration of mononuclear immune cells (monocytes, macrophages, lymphocytes), tissue destruction, and attempts at healing, which include angiogenesis and fibrosis (Bengmark 2004). In chronically inflamed tissue the stimulus is persistent, and therefore recruitment of monocytes is maintained. Existing macrophages are tethered in place, and the proliferation of macrophages is stimulated (especially in atheromatous plaques).

In recent years several biological activities of polyphenols have been described to prevent progression of cancer, cardiovascular diseases, diabetes, the establishment of degenerative diseases, obesity and ageing.

Besides their anti-oxidant properties, these compounds are able to modify physiological and pathological conditions by distinct manners. Epidemiological and experimental studies have been focused on the anti-inflammatory activity of natural polyphenols (Biesalski 2007; Rahman et al. 2006; Santangelo et al. 2007), showing that they may also modulate cellular signalling processes during inflammation or may themselves serve as signalling agents.

The described molecular mechanisms related to the anti-inflammatory properties of polyphenols can involve: (a) inhibition of enzymes of the arachidonic acid pathway, such as cyclooxygenases (COX) and lipoxygenases (LOX); (b) inhibition of nitric oxide synthase (NOS); (c) inhibition of nuclear factor-κappa B (NF-κB); (d) regulation of mitogen-activated protein kinase (MAPK); (e) activation of nonsteroidal anti-inflammatory drugs (NSAID)-activated gene-1 (NAG-1) and (f) regulation of cytokines production (Biesalski 2007; Issa et al. 2006; Santangelo et al. 2007).

8.5.1 Inhibition of Enzymes of Arachidonic Acid Pathway

Arachidonic acid (AA) is a fatty acid released from phospholipid layers of cellular membranes by the enzyme phospholipase $A_2(PLA_2)$. AA is then metabolized by the COX pathway into PG and thromboxane $A_2(TXA_2)$, or by LOX pathway to hydroperoxyeicosatetraenoic acids (HETE) and LT. Two isoforms of COX have been reported, COX-1, constitutively expressed in many tissues, and COX-2, an inducible enzyme expressed in inflammatory related cells after stimulation by pro-inflammatory cytokines or lipopolysaccharide (LPS), and that can produce large amounts of PG. COX-2 is usually not detected in normal tissues. One variant form, COX-3, has recently been reported as well (Chandrasekharan et al. 2002). LOXs have been found in several cells and tissues. The 5-LOX and 12-LOX produce 5-HETE and 12-HETE, respectively, inducers of inflammatory response, while 15-LOX synthesizes the anti-inflammatory 15-HETE (Issa et al. 2006). Several studies demonstrated that polyphenols are able to inhibit cellular enzymes like PLA_2, COX and LOX, reducing AA, PG and LT production, and exhibiting anti-inflammatory actions. The inhibition of PLA_2 has been investigated in order to block the production of inflammatory modulators resulting from AA metabolism. Kim and collaborators tested several flavonoids in the skin and observed inhibition of PLA_2 activity, although the effects on AA metabolism differ between the tested compounds (Kim et al. 2001). Curcumin, a polyphenol derived from tumeric plant, also affects AA metabolism by blocking the phosphorylation of PLA_2, decreasing the expression of COX-2 and inhibiting the catalytic activities of 5-LOX (Hong et al. 2004). These results support the idea that polyphenols might simultaneously affect multiple targets in one or more signalling pathways (Issa et al. 2006).

A battery of polyphenols seems to inhibit COX at either transcriptional or pos-transcriptional levels. Resveratrol, a polyphenol from grapes and wine, inhibited the expression and activity of COX-2 in multiple cancer cell lines (Luceri et al. 2002) and the same effect was observed for black tea polyphenols. Green tea extract and epigallocatechin-3-gallate (EGCG), a polyphenol present in green tea, downregulated COX-2 in stimulated human mammary epithelial cells (Kundu et al. 2003), by decreasing the activation of extracellular signal-regulated protein kinase (ERK) and p38 MAPK, which are upstream enzymes known to regulate COX-2 expression in many cell types. In LPS-activated macrophage RAW264 cells, green tea polyphenols suppressed COX-2 mRNA and protein expression and, therefore, PGE_2. This inhibitory effect occurs through the downregulation of NF-κB and MAPK pathways (Hou et al. 2007). Quercetin (a polyphenol abundant in apples, onions, vegetables, olive oil, fruits, grapes, red wine and tea) also inhibited COX-2 and inducible NOS (iNOS) expression in stimulated RAW264.7 macrophages, probably through NF-κB pathway (De Stefano et al. 2007).

Several polyphenols such as curcumin, quercetin, kaempferol and myricetin, inhibited 5-LOX or 12-LOX activities *in vitro* (Laughton et al. 1991, Yoon and Baek 2005). Proanthocyanidins were also found to be potent inhibitors of 5-LOX. The LOX pathways produce LT, potent mediators of the inflammatory process. When only COX pathways are blocked, LOX still produce inflammatory mediators. Dual inhibition of both COX/LOX pathways seems to be an interesting target. Curcumin is able to block both pathways as previously described (Hong et al. 2004).

8.5.2 Inhibition of Nitric Oxide Synthase Enzymes

Physiological levels of NO, essential to maintain normal body function, are produced by constitutively expressed endothelial NOS (eNOS) and neuronal NOS (nNOS), while iNOS is responsible for prolonged production of higher amounts of NO. iNOS is induced by bacterial products, oxidative stress and inflammatory cytokines in macrophages and other types of cells (Alderton et al. 2001). NO production is increased in inflammation and has pro-inflammatory effects. NO also acts as a second messenger by inducing the production of angiogenic factors, such as vascular endothelial growth factor (VEGF) (Cianchi et al. 2004). NO may also induce tissue damage by reacting with superoxide radical and producing peroxinitrite anion (Cianchi et al. 2004, Issa et al. 2006).

Hämäläinen and collaborators (2007) systematically compared, in standardized experimental conditions, a battery of flavonoids and related compounds belonging to eight different classes (flavones, isoflavones, flavonols, flavanones, flavanols, anthocyanins, hydroxybenzoic and hydroxycinamic acids) on the induction of iNOS expression and NO production in activated macrophages. Flavonols and isoflavones were the most effective compounds. Eight from the 36 compounds tested (flavone, the isoflavones daidzein and genistein, the flavonols isorhamnetin, kaempferol and quercetin, the flavanone naringenin, and the anthocyanin pelargonidin) blocked iNOS protein and mRNA expression and also NO production in a dose-dependent manner. This effect seems to result in part from the inhibition of NF-κB, which acts as a transcription factor for iNOS. Genistein, kaempferol, quercetin, and daidzein also inhibited the activation of the signal transducer and activator of transcription 1 (STAT-1), another important transcription factor for iNOS. The most potent inhibitors of iNOS expression and NO production, genistein, kaempferol and quercetin, inhibited both NF-κB and STAT-1 activation, whereas flavonoids that prevented only NF-κB had a smaller effect on iNOS expression, implying a synergistic but independent effect of those two pathways. Both NF-κB and STAT-1 are involved in the regulation of several other inflammatory genes. For this reason, compounds that inhibit both transcription factors are more effective in down-regulating not only NO but also other inflammatory mediators. Kim et al. (2005) have already observed that genistein and genistin suppressed the DNA binding activation of NF-κB, reducing NO production and iNOS expression in LPS-activated RAW264.7 cells, while treatments with kaempferol and quercetin resulted in similar effects on NO production but by inhibiting the activation of transcription factor activator protein-1 (AP-1). Reduction of NO production and iNOS expression in LPS-primed J774.A1 cells by kaempferol was further described (Autore et al., 2001). Altogether, these findings suggest that polyphenols inhibit NO release by suppressing iNOS expression and/or activity and that NO reduction is frequently mediated by NF-κB inhibition (Biesalski 2007; Issa et al. 2006; Santangelo et al. 2007; Yoon and Baek, 2005).

The anti-inflammatory effects of tea catechins may result from the modulation of the three different NOS isoforms (Santangelo et al. 2007). EGCG and other catechins inhibited the induction of iNOS mRNA and activity in rodent cell lines after stimulation with LPS or interferon γ (IFNγ) (Chan et al. 1997; Paquay et al. 2000), apparently by preventing binding of NF-κB to the promoter of the iNOS gene. EGCG exert its effect on iNOS activity also by competitively inhibiting the binding of arginine and tetrahydrobiopterin, in a process where gallate group seems important (Chan et al. 1997). Treatments of rat aortic rings with EGCG induced vasorelaxation, simultaneously with induction of eNOS activity in endothelial cells

(EC). The NO produced can activate guanylate cyclase to produce cyclic guanosine monophosphate and cause vasorelaxation by phosphoinositide-3 kinase (PI3K), protein kinase A and Akt-dependent signalling pathways (Lorenz et al. 2004). Increased eNOS expression may compensate the endothelial dysfunction, harmonize blood pressure, and prevent atherosclerosis in a long term diet. EGCG has also neuroprotective effects as its administration to Wistar rats before ischemia-induced brain damage, significantly increased eNOS and nNOS (Santangelo et al. 2007).

8.5.3 Impairment of Nuclear Factor-κB

NF-κB is a key regulator of the inflammatory process also present in stress, proliferative and apoptotic cell responses to several different stimuli. NF-κB controls the expression of more than 200 genes encoding pro-inflammatory cytokines (e.g., IL-1, IL-2, IL-6 and TNFα), chemokines (e.g., IL-8, MIP-1α and MCP1), adhesion molecules (e.g., ICAM, VCAM and E-selectin), acute-phase proteins, immune receptors, growth factors (VEGF) and inducible enzymes (COX-2, matrix metalloproteinases (MMP), iNOS, all involved in inflammation as well as in angiogenesis, cell proliferation, adhesion, migration and invasion (Santangelo et al. 2007). Thus, compounds that may suppress NF-κB may have the potencial to prevent, delay or treat inflammatory diseases. The NF-κB/Rel family consists of five members: p65 (RelA), RelB, c-Rel, p50/p105 (NF-κB1) and p50/p100 (NF-κB2), all of them having a conserved region called the Rel homology region. NF-κB function as a dimmer, being the most common the heterodimer formed by p50 and p65 subunits, but other variants can also occur (Vanden Berghe et al. 2006). Homodimmers of p50 and p52 are, in general, repressors of κB site transcription. In unstimulated cells, NF-κB2 is sequestered in the cytoplasm as an inactive, non-DNA-binding form, associated to the inhibitor κB proteins (IκB). In the presence of a stimuli, IκB proteins are rapidly phosphorylated by IκB kinase (IKK) complex, which targets the inhibitor proteins for ubiquitination and subsequent degradation by ubiquitin-proteosome pathway. The NF-κB dimmer released, can then translocate into the nucleous, inducing the expression of various genes (Hayden and Ghosh 2004; Santangelo et al. 2007). While extracellular stimuli persists, continuous IκB degradation maintains NF-κB activity. As NF-κB is present in the cell in an inactive state and do not require new protein synthesis to be activated, it acts as a rapid responder to harmful cellular stimuli (Gilmore 2006).

Polyphenols may exert their anti-inflammatory activity by modulating NF-κB at multiple steps. As a redox-sensitive transcription factor, it is in part regulated by the redox status of the cell. A poor anti-oxidant status favours NF-κB overexpression. Cells of monocyte/macrophage lineage play an important role in response to inflammation. Activation of macrophages by LPS induces them to secrete pro-inflammatory cytokines such as IL-1β, TNFα, NO, PG and COX. LPS-induced NF-κB activation may be inhibited by resveratrol (Djoko et al. 2007). Resveratrol significantly reduced the activity of NF-κB p50 and p65 subunits in vascular smooth muscle cells (VSMC) (Soares and Azevedo 2007). It had no effect on the binding of NF-κB to the DNA, but it blocked the TNFα-induced translocation of p65 subunit of NF-κB to the nucleus and reporter gene transcription (Manna et al. 2000). Polyphenols from grape seeds were able to inhibit nuclear p65 NF-κB and IκB mRNA production in activated RAW 264.7 macrophages (Terra et al. 2007). Through the inhibition of NF-κB, resveratrol suppressed the expression of VCAM-1 mRNA and protein. By inhibiting IκB degradation and, thus, preventing NF-κB binding to DNA, EGCG diminished IL-12 p40 production and iNOS expression in activated-macrophages (Lin and Lin 1997). EGCG action on IKK may be direct or through the IKK- IκB interaction, and the gallate group seems to be important for this effect. IKK regulation appears to be a key point in NF-κB pathway and so, an interesting target. Catechin and epicatechin also decreased NF-κB activity. Expression of IL-8, an important neutrophil chemoattractant and inflammatory mediator, depends on IL-1β activation of NF-κB. EGCG markedly inhibited IL-1β-mediated IL-1β receptor-associated kinase (IRAK) degradation and the downstream signalling events from IRAK degradation: IKK activation, IκB degradation, and NF-κB activation (Wheeler et al. 2004).

AP-1, like NF-κB, is a redox sensitive transcription factor that regulates the transcription of genes involved in cell cycle progression and proliferation. They act independently or co-ordinately to regulate the expression of several target genes, including COX-2 and iNOS. It is noteworthy that the regulation of these two transcription factors is often through the modulation of their activities rather than their expression (Surh 2003). Curcumin, EGCG and resveratrol suppress NF-κB and AP-1 activation *in vitro* (Katiyar et al. 2001; Surh et al. 2001; Yu et al. 2001). Paradoxically, the addition of EGCG, quercetin and gallic acid to cell culture media leads to the generation of substantial amounts of hydrogen peroxide (H_2O_2) (Long et al. 2000; Loo 2003; Rahman et al. 2006). Therefore, polyphenols can either scavenge the constitutive H_2O_2or paradoxically generate additional amounts of H_2O_2 to inhibit the proliferation of cancer cells (Issa et al. 2006). Through the inhibition of NF-κB activation, quercetin diminished IL-8 and MCP-1 production by TNFα stimulated cells (Sato et al. 1997) and also IL-1β and IL-6. This effect is probably due to the inhibition of IκB phosphorylation and the nuclear translocation of p65 subunit. It also blocks NF-κB binding to DNA, in a dose dependent manner. Quercetin inhibited the recruitment of the NF-κB cofactor CBP/p300, a hystone acetyltransferase, to several gene promotors, suggesting it may specifically affect chromatin remodelling of pro-inflammatory genes in murine intestinal cells (Ruiz et al. 2007). Curcumin mediated suppression of NF-κB was attributed to suppression of IKK activity and subsequent IκB phosphorylation and degradation (Jobin et al. 1999). Curcumin also inhibits TNFα-induced NF-κB dependent expression of reporting genes and blocked cigarette smoke-mediated induction of NF-κB binding to DNA, through IKK inactivation and the subsequent steps already described (Jobin et al. 1999; Rahman et al. 2006).

Thus, the modulation of NF-κB pathway by polyphenols can occur at early stages by regulation of the redox state and IKK activation, but also at late stages, such as the binding of NF-κB to DNA. Polyphenol effects on the NF-κB pathway may be one of the reasons why these compounds provide health benefits.

8.5.4 Regulation of Mitogen-Activated Protein Kinase

Nevertheless, NF-κB requires assistance from other transducing pathways like MAPK. In recent years, phenolic compounds were shown to modulate the MAPK pathway, acting in several steps of the cascade and downstream effectors (Soobrattee et al. 2005).

Several polyphenols inhibited ERK, JNK and p38 activities, preventing TNFα-stimulated ICAM-1 expression in respiratory epithelial cells depending on the cell types (Chen et al. 2004; Xagorari et al. 2002). Curcumin and green tea polyphenols modulate several kinase signalling pathways such as JNK, p39, AKT, JAK, ERK and PKC in different cell types (Duvoix et al. 2005; Chen et al. 2000; Huang et al. 2006; Kundu and Surh 2007).

8.5.5 Activation of Nonsteroidal Anti-Inflammatory Drugs-Activated Gene-1

NSAID-activated gene-1 (NAG-1), a member of the transforming growth factor β (TGFβ) superfamily is induced by NSAID and also by dietary compounds. The induction of NAG-1 by NSAID is an important mechanism by which some of the proven anti-inflammatory compounds mediate their effects. NAG-1 protein has a broad activity in inflammation, cancer and differentiation. Indeed, NAG-1 reduces TNFα secretion in macrophages and diminishes proliferation of primitive hemaetopoietic progenitors and several epithelial cell lines (Baek et al. 2004; Biesalski 2007; Yoon and Baek 2005).

8.5.6 Regulation of Cytokines Production

Cytokines are important cell mediators required for an integrated and efficient response in the presence of stimuli, during immune and inflammatory processes. They have been identified in many different tissues and inflammatory diseases. The response results from a balance between the effects of pro-inflammatory cytokines (IL-1β, IL-2, TNFα, IL-6, IL-8 and IFNγ) and anti-inflammatory cytokines (IL-10, IL-4, TGFβ), determining the outcome of the inflammation state and the disease (Santangelo et al. 2007).

Bone marrow-derived macrophages stimulated with a wide variety of polyphenols exhibit reduced TNFα and stimulated the anti-inflammatory cytokine IL-10 expression (Comalada et al. 2006). Tea catechins, epicatechin-3-gallate, epigallocatechin and EGCG decreased the production of IL-1β and enhanced the production of IL-10, but had no effects on the production of IL-6 and TNFα in peripheral blood mononuclear cells (Crouvezier et al. 2001). Kaempferol dose-dependently inhibited IFNγ in murine spleen cells as well as in T cells (Okamoto et al. 2002a). Studying a rodent model, Birrell found out that resveratrol possesses anti-inflammatory properties as it caused a reduction in TNFα, IL-1β and myeloperoxidase (MPO), a marker of neutrophyl activation (Birrel et al. 2005). Contrary to results previously obtained, this effect did not seem to involve NF-κB regulation. Resveratrol also inhibited the production of IFNγ and IL-2, by spleen lymphocytes, and the production of TNFα and IL-12 by peritoneal macrophages (Gao et al. 2001). These results strongly support the idea that polyphenols have the capacity to modulate the immune response and have a potential anti-inflammatory activity, probably due to a balance between pro- and anti-inflammatory cytokine expression, specific for some cytokines and influenced by polyphenols structures (Santangelo et al. 2007).

Generally, a single polyphenol is likely to affect more than one cell process. The effects may be synergistic, additive or even paradoxical. The ability to prevent cross talks between different signalling pathways might be important for the anti-inflammatory properties characteristic of the polyphenols. They may be time or concentration-dependent, cell type specific or non-specific. Therefore, the overall effect should be carefully interpreted.

8.6 Angiogenesis and Polyphenols

During the angiogenic process, new blood vessels develop from the already existent microvascular bed. In response to hypoxia, injured or inflammed tissues, angiogenic factors are released, bind to their receptors and activate EC (reviewed in Chapter 5). The anti-angiogenic effect of plant-derived natural polyphenols has been a matter of study for the last years.

8.6.1 Inhibition of Matrix Metalloproteinases and VEGF Expression

Matrix metalloproteinases (MMP)-2 and MMP-9 have been identified as major MMPs expressed in vascular tissues. They degrade basement membrane, extracelular matrix components, promoting EC and VSMC migration and contributing to angiogenesis.

Red wine and green tea polyphenols diminished VSMC invasion induced by platelet-derived growth factor $(PDGF)_{BB}$ and thrombin, by preventing MMP-2 activation, through the inhibition of catalytic activity of membrane type 1-MMP (MT1-MMP) in a reversibly manner (revised by Oak et al. 2005). Resveratrol and EGCG inhibited MMP-2 and MMP-9 expression in several cell types (Mojzis et al. 2008) and, the anti-angiogenic effects of quercetin may also be related to an inhibition of the expression and activity of MMP-2 and MMP-9. Abrogation of MMP-9 gene expression is a key mediator for the anti-angiogenic effects exerted by curcumin and its analogs (Kim et al. 2002).

Red wine polyphenols strongly inhibited growth factor-induced VEGF expression in VSMC at concentrations likely to be achieved in human blood after moderate red wine consumption (Oak et al. 2003; Oak et al. 2005), partly by the selective prevention of the redox-sensitive activation of the p38-MAPK pathway. Resveratrol as well as anthocyanins downregulated the production of several angiogenic cytokines, including VEGF and IL-8 (Dulak 2005; Lamy et al. 2006). Green tea polyphenols and EGCG were also shown to decrease VEGF production in head and breast carcinoma cells, by inhibiting epidermal growth factor receptor pathways such as the constitutive activation of Stat3 and NF-κB (Masuda et al. 2002). EGCG significantly blocked hypoxia- and serum-induced hypoxia-induced factor-1α (HIF-1α) protein accumulation in human cervical carcinoma (HeLa) and hepatoma (HepG2) cancer cells, resulting in decreased VEGF expression at mRNA and protein levels. This mechanism seems to involve blocking of both PI3K/Akt and Erk1/2 pathways and enhancement of HIF-1α degradation by proteosoma (Zhang et al. 2006). EGCG at physiological concentrations inhibited VEGF binding to its receptor, decreasing PI3K activity and NF-κB binding to DNA. Genistein, a soy isoflavone, suppressed VEGF and fibroblast growth factor (FGF)-2 expression (Fan et al. 2006). Curcumin was found to completely prevent VEGF synthesis by microvascular EC stimulated with advanced glycation end products in a process mediated by the downregulation of NF-κB and AP-1 activity (Okamoto et al. 2002b).

8.6.2 In Vitro *and* In Vivo *Anti-Angiogenic Effects*

EC proliferation occurs early in angiogenesis and continues as the new capillary elongates. Activation of PI3K/Akt promotes EC survival and proliferation. MAPK signalling pathways (ERK1/2, p38, JNK) mediate growth factor and mechanical force-induced proliferation (Mojzis et al. 2008). Several studies indicated that polyphenols are able to inhibit proliferation and migration of vascular wall cells.

Resveratrol prevented S:G2 cell cycle phases progression in EC, accompanied by an increase in expression of p53 protein and of the cyclin-dependent kinase inhibitor p21 (Hsieh et al. 1999). Several transcription factors such as NF-κB, AP-1, and Erg-1 and the expression of anti-apoptotic genes were downregulated (reviewed by Dulak 2005). Red wine and green tea polyphenolic compounds are able to inhibit several key events of the angiogenic process like the migration and proliferation of EC and VSMC, by specific inhibition of p38 MAPK and PI3K/Akt pathways (Iijima et al. 2002; Oak et al. 2005). Red grape skin polyphenols extract inhibited platelet-derived lipid sphingosine-1-phosphate (S1P) and VEGF-mediated EC chemotaxis (Barthomeuf et al. 2006). Soares and Azevedo additionally suggested that NF-κB is a putative downstream effector of S1P, controlling angiogenesis by these polyphenols (2007). In addition, EGCG induced EC apoptosis through mitochondrial depolarization and activation of caspase-3, and also VSMC apoptosis in a p53- and NF-κB-dependent manner, reducing vascular cells proliferation. EGCG also inhibited VEGF-induced EC proliferation, migration and tube formation by blocking VEGF binding to its receptor and by decreasing VEGFR-1 and -2 autophosphorylation (revised in Stangl et al. 2007). EC proliferation, migration and tube formation, three main steps of the angiogenic process, are prevented by several polyphenolic compounds (Mojzis et al. 2008; Singh et al. 2006; Bertl et al. 2004; Albini et al. 2006). Negrão et al. reported that 10 μM XN, a chalcone present in hops, significantly decreased viability and invasion capacity and increased apoptosis in EC as well as in smooth muscle cells (2007), leading to a drastic decrease in the number of capillary-like structures formed by culturing EC on Matrigel. These authors further proposed that NF-κB signalling inactivation may be one of the pathways triggered by this compound in both cells types (Negrão et al. 2007b). Another beer polyphenol (isoxanthohumol, IXN) had similar anti-angiogenic properties as it diminished cell viability, migration, invasion and capillary-like structures formation, while increased apoptosis. Surprisingly, 8-prenylnaringenin, another hop polyphenol, seemed to have pro-angiogenic effects, as it increased HUVEC and HASMC viability, migration and invasion, diminished cell apoptosis and increased capillary-like structures formation (Negrão et al. 2007a). In fact, there is some evidence that polyphenols may also exert pro-angiogenic effects. Grape seed proanthocyanidin extract containing resveratrol potentiate wound angiogenesis by enhancing the oxidizing environment and thus, stimulating VEGF production (Khanna et al. 2002).

Evaluation of the *in vivo* polyphenols efficacy in the angiogenic process are less common, and most of them refer to angiogenesis related to cancer development. Recently, Baron-Menguy and collaborators highlighted a dose-dependent effect of red wine polyphenols in an *in vivo* model of ischemia. These molecules have a unique dual effect and offer important therapeutic perpectives for prevention and treatment of ischemic diseases (low doses, pro-angiogenic effect) and cancer (high doses, anti-angiogenic effect), through the mediation of PI3K-Akt-eNOS pathways and potential MMP modulation (Baron-Menguy et al. 2007). Resveratrol may inhibit tumour-induced neovascularisation *in vivo* due to blockade of both VEGF and bFGF-receptor mediated responses, involving the MAPK phosphorylation (Brakenhielm et al. 2001). In accordance with the role of angiogenesis in tumour growth, oral administration of resveratrol inhibited the growth of murine fibrosarcoma in mice and delayed wound healing.

Local application of red wine and green tea polyphenols into the chick embryo choroallantoic membrane (CAM), strongly inhibited angiogenesis. Green tea polyphenols seem to diminish in vivo tumour growth, microvessel density and tumour cell proliferation (Mojzis et al. 2008), suggesting that EGCG may exert at least part of its anti-cancer effect by inhibiting angiogenesis and blocking VEGF.

Infusions of green tea polyphenols reduced angiogenesis and metastasis markers, especially VEGF, MMP and urokinase plasminogen activator in a xenotransplant model of prostate cancer, CAM assay, and corneal neovascularisation in mice (Adhami et al. 2003; Cao and Cao 1999; Oak et al. 2005; Dulak 2005). When administered to mice in the drinking water green tea or purified EGCG inhibited angiogenesis in the Matrigel plug assay and restrained Kaposi's sarcoma tumour growth, as confirmed by histological analysis (Fassina et al. 2004).

XN effectively blocked tumour angiogenesis and tumour growth *in vivo*. Subcutaneous application of XN leads to inhibition of the growth of the breast tumour xenografts and tumour-induced neovascularisation (Gerhäuser 2005). Oral administration of XN to nude mice inoculated with MCF7 cells resulted in reduced inflammation, increased the percentage of apoptotic cells and decreased microvessel density, indicating that XN is simultaneously an anti-angiogenic and anti-cancer compound (Monteiro et al. 2008).

Although experimental models have shown in vivo anti-angiogenic effects by several polyphenols, to date no clinical research was able to prove the efficacy of these compounds in humans. Moreover, toxicity studies with pharmacological doses of polyphenolic compounds are mandatory in order to determine their potential usefulness in disease prevention and treatment. Meanwhile, angiogenesis can be overcome with a systemic non-toxic low dose of angiogenesis inhibitors over a long period of time providing anti-angiogenic foods a good strategy for the prevention of degenerative diseases (reviewed by Fan et al. 2006).

8.7 Polyphenols and Metabolic Syndrome

The metabolic syndrome has become increasingly common in western countries and the dominant underlying risk factors for this syndrome appear to be abdominal obesity and insulin resistance (Miranda et al. 2005). We will focus on obesity, type 2 diabetes mellitus (T2DM) and atherosclerosis, three of the most relevant disorders associated with the metabolic syndrome and the role of polyphenols in the improvement of these conditions.

8.7.1 Obesity

Obesity is now recognized as a state of chronic, low-grade inflammation (Lyon et al. 2003). In fact, adipocytes secrete hormones, proteins and adipokines, which among other effects are able to promote inflammation and angiogenesis (Soares and Azevedo 2007). Furthermore, obese adipose tissue (AT) is characterized by an enhanced infiltration of macrophages that produce various local pro-inflammatory mediators (like TNF-α, MCP-1 and NO) (Weisberg et al. 2003), which are also released into the systemic circulation. Recent reports suggest that angiogenesis is an active process in AT, necessary for the development of obesity. Therefore, angiogenesis inhibition by polyphenols would be a good therapeutic strategy on obesity control.

As revised by Cheng (2006), tea may be helpful in obesity control, either through stimulation of hepatic lipid metabolism, lipase inhibition, thermogenesis stimulation, modulation of appetite or synergism with caffeine. EGCG presented beneficial effects on human health, by reducing adipocyte differentiation and decreasing triglyceride levels. In a recent clinical trial, green tea extract containing 25% EGCG reduced body weight and waist circumference in moderately obese patients. However, epidemiological evidence regarding the effects of tea consumption on obesity-related disease is still controversial (Moon et al. 2007). Long term consumption of green tea may decrease the incidence of obesity and perhaps green tea compounds like EGCG may be useful for the treatment of obesity. The unique therapeutic advantages for anthocyanins in regulation of adipocyte function has been evidenced recently by Tsuda (2008), including adipocytokines expression with important implications for prevention of obesity and metabolic syndrome (MS).

In fact, mice fed during 12 weeks with a high-fat diet containing anthocyanin dysplayed reduced weight, liver triacylglycerol, serum insulin and glucose concentrations. Incubation of LPS-stimulated RAW 264 macrophages cocultured with 3T3-L1 adipocytes with naringenin led to inhibition of TNFα, MCP-1, and NO, reinforcing the effect of naringenin in ameliorating the inflammatory changes in obese AT (Hirai et al. 2007).

8.7.2 Diabetes

A characteristic feature of obese patients is the development of insulin resistance and consequently T2DM. Diabetes is a pathology in which inflammation and angiogenesis are interrelated. Accordingly, besides hyperglycaemia and increased insulin levels, diabetes mellitus also exhibit enhanced angiogenesis associated with diabetic retinopathy, nephropathy and potentially instability of atherosclerotic plaques. Simultaneously, defective vascularization is observed elsewhere, leading to wound healing impairment and defective collateral growth in coronary heart disease (Soares and Azevedo 2007).

Curcumin treatment significantly reduced macrophage infiltration of white adipose tissue, increased AT adiponectin production, and decreased hepatic NF-κB activity (Weisberg et al. 2008), resulting in ameliorated diabetes in obese and leptin-deficient ob/ob male C57BL/6J mice, as determined by glucose and insulin tolerance testing and hemoglobin A1c percentages (reviewed by Opie and Lecour 2007). Green tea infusion increased insulin sensitivity by enhancing glucose uptake and insulin binding to adipocytes in Sprague-Dawley rats (Wu et al. 2004; reviewed by Cheng 2006). The recent discovery of an EGCG receptor localized in several cell types may in part explain the numerous biological effects observed for EGCG (Kao et al. 2006). Cinnamon polyphenols also improve insulin sensitivity by reducing mean fasting serum glucose, triglyceride, total cholesterol and low density lipoprotein-cholesterol in subjects with T2DM. Patients with MS also improved fasting blood glucose, systolic blood pressure and percentage of body fat, with cinnamon aqueous extract consumption (Anderson 2008). A polyphenol-rich Aloe vera extract (350 mg/kg) administered orally to insulin resistant ICR mice, was able to decrease significantly both body weight and blood glucose levels, suggesting that Aloe vera could be effective in controling insulin resistance (Pérez et al. 2007). In addition, quercetin improved rat diabetic status in terms of urine volume, urine sugar, and fasting blood glucose (Shetty et al. 2004). Anthocyanins from different sources have been shown to affect glucose absorption, insulin level, secretion and action and lipid metabolism *in vitro* and *in vivo*. Extracts from blue-berry increase glucose uptake and protect neural cells from the toxic effects of high glucose levels. Anthocyanins decrease fasting glucose and serum cholesterol and decreases hemoglobin A1c in T2DM patients (reviewed by Dembinska-Kiec et al. 2008).

Conversely, other studies could not support the hypothesis that high intake of flavonoids protects against the development of T2DM (Song et al. 2005). Further studies, especially in humans, are needed to verify the potential protection of polyphenols or polyphenol-rich food against the development of T2DM or in ameliorating its complications.

8.7.3 Atherosclerosis

A major consequence of diabetes mellitus is atherosclerosis. Chronic inflammation also plays a key role in atherogenesis. Inflammation of the vessel wall, activation of the vascular endothelium, increased adhesion of mononuclear cells to the injured endothelial layer, and their subsequent extravasations into the vessel wall, proliferation and migration of VSMC, are initial events in the process (Libby 2002). The role and causality of angiogenesis in the development of cardiovascular disease is still a topic of controversy. Nevertheless, the prevalence of neovascularisation in atherosclerotic plaques has been positively correlated with their instability, as the density of the microvessels increased in ruptured plaques and in lesions with macrophages infiltration (Moreno et al. 2004). These microvessels contribute to the progression of coronary atherosclerosis providing oxygen and nutrients and possibly inducing intimal haemorrhage and plaque rupture, since these new blood vessels are fragile. Moreover VEGF is strongly expressed in human atherosclerotic plaques (Chen et al. 1999) and increases gradually with the progression of lesions. VEGF also stimulates the expression of pro-inflammatory and pro-thrombotic molecules in atherosclerotic plaques (Oak et al. 2005).

Epidemiological evidence indicates the existence of a negative correlation between consumption of polyphenol rich foods or beverages and the incidence of cardiovascular disease and stroke (reviewed by Stoclet et al. 2004). In a very recent publication, analyses of five flavonoid subclasses suggested that a high intake of flavonoids, and less strongly flavanols, were associated with decreased risk of ischaemic stroke and possibly cardiovascular diseases (Mursu et al. 2008). Given the relevance of the inflammatory and angiogenic processes in atherosclerosis, the use of anti-inflammatory and anti-angiogenic agents is a very attractive approach. Accordingly, Oak et al. (2005) suggested that red wine and green tea polyphenols have *in vitro* and *in vivo* anti-angiogenic properties by inhibiting the expression of two strong pro-angiogenic factors, VEGF and MMP-2, and also by preventing the proliferation and migration of EC and VSMC, at concentrations likely to be reached in blood after consumption of red wine and green tea. The anti-angiogenic properties of these polyphenols explain the reduced risk of coronary heart diseases following chronic consumption of moderate amounts of red wine and green tea, together with their ability to increase the levels of high density lipoprotein, diminished low-density lipoprotein oxidation, prevent activation of platelets and expression of pro-thrombotic and pro-atherosclerotic molecules as MCP-1. Resveratrol is one of the most active red wine polyphenols respecting to cardiovascular endpoints. Enhanced NO released and improvement of human endothelial function was observed after procyanidin consumption. Anti-oxidant activity should also be noted as underlying mechanism of polyphenols health impact in the vascular system, as atherosclerosis is closely related to oxidative events such as oxidized LDL accumulation in the macrophages. Accordingly, grapeseed extract has also been shown to be a more potent scavenger of oxygen-free radicals than other common anti-oxidants such as vitamin C and E. Dietary polyphenols may also prevent the development of atherosclerosis by inhibiting proliferation and migration of EC and VSMC, important processes in the establishment of atherosclerosis (Oak et al. 2005, Stoclet et al. 2004; Vinson et al. 2001). However, red wine was unable to reduce mature atherosclerotic plaques in apoE-deficient mice (Bentzon et al. 2001). The beneficial effects of dietary polyphenols on vascular ischemic obstruction events might also be related to prevention of thrombosis resulting from decreased platelet activation or from decreased expression of pro-thrombotic and pro-atherosclerotic molecules (Stoclet et al. 2004). Terao group (Kawai et al. 2008), studied the target sites of quercetin underlying its atherosclerotic protective mechanism *in vivo*. They raised a novel monoclonal antibody 14A2 targeting the quercetin-3-glucuronide and found that the activated macrophage might be a potential target of dietary flavonoids in the aorta. Immunohistochemical studies demonstrated that the positive staining specifically accumulates in human atherosclerotic lesions, not in the normal aorta, and that the intense staining was associated with the macrophage-derived foam cells. Resveratrol is a potent polyunsaturated fatty acids (PUFA) oxidation inhibitor (Miller and Rice-Evans 1995), but *in vivo* experiments did not always confirm the anti-atherogenic properties of this polyphenol. Accordingly, resveratrol promoted atherosclerotic development, rather than exerting a protective effect in hyperlipidaemic rabbits (Wilson et al. 1996). Ivanov and collaborators investigated the effects of quercetin and green tea extract on monocyte-binding properties of extracellular matrix (ECM) produced by human aortic endothelial cells. They observed a reduction of monocyte adhesion to endothelium that seemed partly mediated through specific modulation of ECM composition and properties (Ivanov et al. 2008). Chun and collaborators demonstrated that intake of dietary flavonoids is inversely associated with serum C-reactive protein (CRP) concentrations, a sensitive risk factor for cardiovascular diseases, in U.S. adults, suggesting that intake of flavonoid-rich foods may help reduce inflammation processes associated to chronic diseases (Chun et al. 2008). Compelling epidemiological, clinical and experimental evidence suggests that compounds contained in green and black tea are associated with beneficial effects in prevention of cardiovascular diseases, particularly atherosclerosis and coronary heart disease. Tea flavonoids seem to improve endothelial function, and reduce blood pressure, oxidative damage, blood cholesterol concentrations, inflammation and risk of thrombosis (Hodgson 2008). In a recent clinical trial with stable coronary heart disease patients, inhibition of IL-6 and CRP was observed after administration of polyphenol-rich olive oil, suggesting that consumption of virgin olive oil could provides beneficial effects in stable coronary heart disease patients (Fitó et al. 2008).

Current evidence suggests that the protection against cardiovascular diseases associated with polyphenol-rich diets results from the addition of a variety of effects produced by different mechanisms and, in some cases, different compounds.

Besides their anti-oxidant effects, they can act improving the endothelium function and inhibiting angiogenesis and migration and proliferation of vascular cells (Stoclet et al. 2004).

The enormous variety of published work demonstrates the potential of polyphenols as therapeutic tools in inflammatory diseases such as obesity, T2DM and cardiovascular diseases. Polyphenolic compounds that are able to suppress oxidative stress, inflammation and diminished angiogenesis, as referred in previous sections, might be proven useful against obesity and metabolic syndrome. Although many epidemiological studies suggested an association between higher anti-oxidants polyphenols intake and human health, recent large scale trials with anti-oxidant supplementation have failed to confirm this protection against cardiovascular mortality and type 2 diabetes. The metabolic changes associated to the metabolic syndrome are complex and metabolic deregulations takes years to manifest in clinical disease. Therefore, any treatment strategy for preventing these diseases is difficult to implement, and failure of some clinical trials may be partly due to the initiation of such therapy after atherosclerosis and diabetes are already well established. In contrast, the benefits found in animal models may result from early initiation of treatments while the pathology is still evolving (Cheng 2006). Nevertheless, the available evidence does not contradict the advice to increase consumption of fruits and vegetables to reduce the risk of cardiovascular disease especially in patients with diabetes (Dembinska-Kiec et al. 2008).

8.8 Conclusion

Successful strategies are required to halt the increasing prevalence of the MS. Lifestyle interventions like weight loss, exercise and a healthy diet, rich in fruit and vegetables, will greatly improve health and delay MS establishment. Nevertheless, long-term maintenance of lifestyle changes by general population is poor. Thus nutrition advice or supplement with phytochemical compounds with proven effects in human health represents an attractive and potentially effective approach to the problem.

The considerable current interest in the possible beneficial health effects of polyphenol compounds resulted in many laboratory studies of these compound properties on culture cells. Most investigation on the beneficial effects of polyphenols, however, was obtained from *in vitro* studies. More detailed investigations are required to extrapolate these results to *in vivo* situations, given the different and even opposite effects of polyphenols *in vivo*. This is particularly relevant as it is known that polyphenols undergo various biochemical transformations, which affect their bioavailability as well as bio-efficacy. Accordingly, some authors strongly defend that achieved *in vivo* plasma and intracellular concentrations of polyphenols after red wine consumption are not enough for the proposed impact on human physiology at least concerning superoxide scavenging and NO production (Huisman et al. 2004), suggesting prudency when interpreting *in vitro* results obtained with high polyphenolic concentrations. Still of concern is the fact that most laboratories use for their studies cancer cell lines, genetically unstable, often with metabolic differences regarding normal tissue cells, which increases difficulty to extrapolate results and gives little information about the effects of polyphenols in other physiological conditions. Plant polyphenols have attracted much attention with regard to human health, due to their apparent low toxicity, limited costs, and broad availability. The potential underlying physiological mechanisms of many of these natural compounds, however, are not completely understood, although they seem to act at multiple molecular levels. The pathophysiology of diabetes and cardiovascular diseases is multifactorial and comprises processes, which appear to be affected by various polyphenols. Their effects may, therefore, be of potential interest in prevention and treatment of these diseases. A number of unresolved questions, however, still persist before polyphenols can be clinically used or recommended as nutritional supplements, particularly concerning dose, specificity, potency, feasibility, as well as short- and long-term side effects in humans. Interactions between intracellular signalling pathways and polyphenols could also have unpredictable outcomes depending on the cell type, the disease studied and the stimulus applied. Although naturally occurring polyphenols are generally considered to be pharmacologically safe, it is necessary to be aware that these compounds can have detrimental effects in the body, depending on the localisation and cell type upon which they are acting.

Acknowledgments Research work supported by European Research Advisory Board, ERAB (EA0641).

References

Adhami VM, Ahmad N, Mukhtar H. *Molecular targets for green tea in prostate cancer prevention.* J Nutr. 2003; 133: 2417S–24S.

Albini A, Dell'Eva R, Vene R, Ferrari N, Buhler DR, Noonan DM, Fassina G. *Mechanisms of the antiangiogenic activity by the hop flavonoid xanthohumol:NF-kappaB and Akt as targets.* FASEB J. 2006; 20: 20527–29.

Alderton WK, Cooper CE, Knowles RG. *Nitric oxide synthases: structure, function and inhibition.* Biochem J. 2001; 357: 593–615.

Amic D, Davidovic-Amic D, Beslo D, Rastija V, Lucic B, Trinajstic N. *SAR and QSAR of the anti-oxidant activity of flavonoids.* Curr Med Chem. 2007; 14: 827–45.

Anderson RA. *Chromium and polyphenols from cinnamon improve insulin sensitivity.* Proc Nutr Soc. 2008; 67: 48–53.

Aquilano K, Baldelli S, Rotilio G, Ciriolo MR. *Role of Nitric Oxide Synthases in Parkinson's Disease: A Review on the Anti-oxidant and Anti-inflammatory Activity of Polyphenols.* Neurochem Res. 2008; 33(12): 2416–26.

Arora A, Nair MG, Strasburg GM. *Structure-activity relationships for anti-oxidant activities of a series of flavonoids in a liposomal system.* Free Radic Biol Med. 1998; 24: 1355–63.

Autore G, Rastrelli L, Lauro MR, Marzocco S, Sorrentino R, Sorrentino U, Pinto A, Aquino R. *Inhibition of nitric oxide synthase expression by a methanolic extract of Crescentia alata and its derived flavonols.* Life Sci. 2001; 70: 523–34.

Baek SJ, Kim JS, Jackson FR, Eling TE, McEntee MF, Lee SH. *Epicatechin gallate-induced expression of NAG-1 is associated with growth inhibition and apoptosis in colon cancer cells.* Carcinogenesis. 2004; 25: 2425–32.

Baron-Menguy C, Bocquet A, Guihot AL, Chappard D, Amiot MJ, Andriantsitohaina R, Loufrani L, Henrion D. *Effects of red wine polyphenols on postischemic neovascularisation model in rats: low doses are proangiogenic, high doses anti-angiogenic.* FASEB J. 2007; 21: 3511–21.

Barthomeuf C, Lamy S, Blanchette M, Boivin D, Gingras D, Béliveau R. *Inhibition of sphingosine-1-phosphate- and vascular endothelial growth factor-induced endothelial cell chemotaxis by red grape skin polyphenols correlates with a decrease in early platelet-activating factor synthesis.* Free Radic Biol Med. 2006; 40: 581–90.

Bengmark S. *Acute and "chronic" phase reaction-a mother of disease.* Clin Nutr. 2004; 23: 1256–66.

Bentzon JF, Skovenborg E, Hansen C, Moller J, de Gaulejac NS, Proch J, Falk E. *Red wine does not reduce mature atherosclerosis in apolipoprotein E- deficient mice.* Circulation 2001; 103: 1681–7.

Bertl E, Becker H, Eicher T, Herhaus C, Kapadia G, Bartsch H, Gerhauser C. *Inhibition of endothelial cell functions by novel potential cancer chemoprotective agents.* Biochem Biophys Res Commun. 2004; 325: 287–95.

Biesalski HK. *Polyphenols and inflammation: basic interactions.* Curr Opin Clin Nutr Metab Care. 2007; 10: 724–8.

Birrell MA, McCluskie K, Wong S, Donnelly LE, Barnes PJ, Belvisi MG. *Resveratrol, an extract of red wine, inhibits lipopolysaccharide induced airway neutrophilia and inflammatory mediators through an NF-kappaB-independent mechanism.* FASEB J. 2005; 19: 840–1.

Bors W, Heller W, Michel C, Saran M. *Radical chemistry of flavonoid anti-oxidants.* Adv Exp Med Biol. 1990; 264: 165–70.

Bors W, Michel C, Schikora S. *Interaction of flavonoids with ascorbate and determination of their univalent redox potentials: a pulse radiolysis study.* Free Radic Biol Med. 1995; 19: 45–52.

Bosetti C, Rossi M, McLaughlin JK, Negri E, Talamini R, Lagiou P, Montella M, Ramazzotti V, Franceschi S, LaVecchia C. *Flavonoids and the risk of renal cell carcinoma.* Cancer Epidemiol Biomarkers Prev. 2007; 16: 98–101.

Bosetti C, Spertini L, Parpinel M, Gnagnarella P, Lagiou P, Negri E, Franceschi S, Montella M, Peterson J, Dwyer J, Giacosa A, La Vecchia C. *Flavonoids and breast cancer risk in Italy.* Cancer Epidemiol Biomarkers Prev. 2005; 14: 805–8.

Brakenhielm E, Cao R, Cao Y. *Supression of angiogenesis, tumour growth, and wound healing by resveratrol, a natural compound in red wine and grapes.* FASEB J. 2001; 15: 1798–800.

Cao G, Sofic E, Prior RL. *Anti-oxidant and pro-oxidant behavior of flavonoids: structure-activity relationships.* Free Radic Biol Med. 1997; 22: 749–60.

Cao Y, Cao R. *Angiogenesis inhibited by drinking tea.* Nature. 1999; 398: 381–381.

Castillo J, Benavente-Garcia O, Lorente J, Alcaraz M, Redondo A, Ortuno A, Del Rio JA. *Anti-oxidant activity and radioprotective effects against chromosomal damage induced in vivo by X-rays of flavan-3-ols (Procyanidins) from grape seeds (Vitis vinifera): comparative study versus other phenolic and organic compounds.* J Agric Food Chem. 2000; 48: 1738–45.

Chan MM, Fong D, Ho CT, Huang HI. *Inhibition of inducible nitric oxide synthase gene expression and enzyme activity by epigallocatechin gallate, a natural product from green tea.* Biochem Pharmacol. 1997; 54: 1281–6.

Chandrasekharan NV, Dai H, Roos KL, Evanson NK, Tomsik J, Elton TS, Simmons DL. *COX-3, a cyclooxygenase-1 variant inhibited by acetaminophen and other analgesic/antipyretic drugs: cloning, structure, and expression.* Proc Natl Acad Sci USA. 2002; 99: 13926–31.

Chen CC, Chow MP, Huang WC, Lin YC, Chang YJ. *Flavonoids inhibit tumor necrosis factor-alpha-induced up-regulation of intercellular adhesion molecule-1 (ICAM-1) in respiratory epithelial cells through activator protein-1 and nuclear factor-kappaB: structure-activity relationships.* Mol Pharmacol. 2004; 66: 683–93.

Chen YX, Nakashima Y, Tanaka K, Shiraishi S, Nakagawa K, Sueishi K. *Immunohistochemical of vascular endothelial growth factor/vascular permeability factor in atherosclerotic intimas of human coronary arteries.* Arterioscler Tromb Vasc Biol. 1999; 19: 131–9.

Chen C, Yu R, Owuor ED, Kong AN. *Activation of antioxidant-response element (ARE), mitogen-activated protein kinases (MAPKs) and caspases by major green tea polyphenol components during cell survival and death.* Arch pharm Res. 2000; 23: 605–12.

Cheng TO. *All teas are not created equal. The Chinese green tea and cardiovascular health.* Int J Cardiol. 2006; 108: 301–8.

Chun OK, Chung SJ, Claycombe KJ, Song WO. *Serum C-reactive protein concentrations are inversely associated with dietary flavonoid intake in U.S. adults.* J Nutr. 2008; 138: 753–60.

Cianchi F, Cortesini C, Fantappiè O, Messerini L, Sardi I, Lasagna N, Perna F, Fabbroni V, Di Felice A, Perigli G, Mazzanti R, Masini E. *Cyclooxygenase-2 activation mediates the proangiogenic effect of nitric oxide in colorectal cancer.* Clin Cancer Res. 2004; 10:2694–704.

Comalada M, Ballester I, Bailón E, Sierra S, Xaus J, Gálvez J, de Medina FS, Zarzuelo A. *Inhibition of pro-inflammatory markers in primary bone marrow-derived mouse macrophages by naturally occurring flavonoids: analysis of the structure-activity relationship.* Biochem Pharmacol. 2006; 72: 1010–21.

Cooper-Driver GA, Bhattacharya M. *Role of phenolics in plant evolution.* Phytochemistry. 1998; 49: 1165–74.

Corder R, Mullen W, Khan NQ, Marks SC, Wood EG, Carrier MJ, Crozier A. *Oenology: red wine procyanidins and vascular health.* Nature. 2006; 444: 566.

Cotelle N. *Role of flavonoids in oxidative stress.* Curr Top Med Chem. 2001; 1: 569–90.

Crouvezier S, Powell B, Keir D, Yaqoob P. *The effects of phenolic components of tea on the production of pro- and anti-inflammatory cytokines by human leukocytes in vitro.* Cytokine. 2001; 13: 280–6.

Cushnie TP, Lamb AJ. *Antimicrobial activity of flavonoids.* Int J Antimicrob Agents. 2005; 26: 343–56.

Day AJ, Canada FJ, Diaz JC, Kroon PA, McLauchlan R, Faulds CB, Plumb GW, Morgan MR, Williamson G. *Dietary flavonoid and isoflavone glycosides are hydrolysed by the lactase site of lactase phlorizin hydrolase.* FEBS Lett. 2000; 468: 166–70.

Dembinska-Kiec A, Mykkanen O, Kiec-Wilk B, Mykkanen H. *Anti-oxidant phytochemicals against type 2 diabetes.* Br J Nutr. 2008; 99: ES109–17.

De Stefano D, Maiuri MC, Simeon V, Grassia G, Soscia A, Cinelli MP, Carnuccio R. *Lycopene, quercetin and tyrosol prevent macrophage activation induced by gliadin and IFN-gamma.* Eur J Pharmacol. 2007; 566: 192–9.

Di Carlo G, Mascolo N, Izzo AA, Capasso F. *Flavonoids: old and new aspects of a class of natural therapeutic drugs.* Life Sci. 1999; 65: 337–53.

Djoko B, Chiou RY, Shee JJ, Liu YW. *Characterization of immunological activities of peanut stilbenoids, arachidin-1, piceatannol, and resveratrol on lipopolysaccharide-induced inflammation of RAW 264.7 macrophages.* J Agric Food Chem. 2007; 55: 2376–83.

Dragoni S, Gee J, Bennett R, Valoti M, Sgaragli G. *Red wine alcohol promotes quercetin absorption and directs its metabolism towards isorhamnetin and tamarixetin in rat intestine in vitro.* Br J Pharmacol. 2006; 147: 765–71.

Dugas AJ, Jr., Castaneda-Acosta J, Bonin GC, Price KL, Fischer NH, Winston GW. *Evaluation of the total peroxyl radical-scavenging capacity of flavonoids: structure-activity relationships.* J Nat Prod. 2000; 63: 327–31.

Dulak J. *Nutraceuticals as anti-angiogenic agents: hopes and reality.* J Physiol Pharmacol. 2005; 56: 51–67.

Duvoix A, Blasius R, Delhalle S, Schnekenburger M, Morceau F, Henry E, Dicato M, Diederich M. *Chemopreventive and therapeutic effects of curcumin.* Cancer Lett. 2005; 223: 181–90.

Fan TP, Yeh JC, Leung KW, Yue PY, Wong RN. *Angiogenesis: from plants to blood vessels.* Trends Pharmacol Sci. 2006; 27: 297–309.

Fassina G, Vene R, Morini M, Minghelli S, Benelli R, Noonan DM, Albini A. *Mechanisms of inhibition of tumor angiogenesis and vascular tumor growth by epigallocatechin-3-gallate.* Clin Cancer Res. 2004; 10: 4865–73.

Fitó M, Cladellas M, de la Torre R, Martí J, Muñoz D, Schröder H, Alcántara M, Pujadas-Bastardes M, Marrugat J, López-Sabater MC, Bruguera J, Covas MI; SOLOS Investigators. *Anti-inflammatory effect of virgin olive oil in stable coronary disease patients: a randomized, crossover, controlled trial.* Eur J Clin Nutr. 2008; 62: 570–4.

Galati G, O'Brien PJ. *Potential toxicity of flavonoids and other dietary phenolics: significance for their chemopreventive and anticancer properties.* Free Radic Biol Med. 2004; 37: 287–303.

Gao X, Xu YX, Janakiraman N, Chapman RA, Gautam SC. *Immunomodulatory activity of resveratrol: suppression of lymphocyte proliferation, development of cell-mediated cytotoxicity, and cytokine production.* Biochem Pharmacol. 2001; 62: 1299–308.

Gee JM, DuPont MS, Day AJ, Plumb GW, Williamson G, Johnson IT. *Intestinal transport of quercetin glycosides in rats involves both deglycosylation and interaction with the hexose transport pathway.* J Nutr. 2000; 130: 2765–71.

Gerhäuser C. *Beer constituents as potential cancer chemopreventive agents.* Eur J Cancer. 2005; 41: 1941–54.

Gilmore TD. *Introduction to NF-B: players, pathways, perspectives.* Oncogene.2006; 25:6680–4.

Gonçalves R, Soares S, Mateus N, De Freitas V. *Inhibition of trypsin by condensed tannins and wine.* J Agric Food Chem. 2007; 55: 7596–601.

Hämäläinen M, Nieminen R, Vuorela P, Heinonen M, Moilanen E. *Anti-inflammatory effects of flavonoids: genistein, kaempferol, quercetin, and daidzein inhibit STAT-1 and NF-kappaB activations, whereas flavone, isorhamnetin, naringenin, and pelargonidin inhibit only NF-kappaB activation along with their inhibitory effect on iNOS expression and NO production in activated macrophages.* Mediators Inflamm. 2007; 2007: 45673.

Hammerstone JF, Lazarus SA, Schmitz HH. *Procyanidin content and variation in some commonly consumed foods.* J Nutr. 2000; 130: 2086S–92S.

Hanasaki Y, Ogawa S, Fukui S. *The correlation between active oxygens scavenging and antioxidative effects of flavonoids.* Free Radic Biol Med. 1994; 16: 845–50.

Hayden MS, Ghosh S. *Signaling to NF-kappaB.* Genes Dev. 2004; 18: 2195–224.

Heim KE, Tagliaferro AR, Bobilya DJ. *Flavonoid anti-oxidants: chemistry, metabolism and structure-activity relationships.* J Nutr Biochem. 2002; 13: 572–84.

Hirai S, Kim YI, Goto T, Kang MS, Yoshimura M, Obata A, Yu R, Kawada T. *Inhibitory effect of naringenin chalcone on inflammatory changes in the interaction between adipocytes and macrophages.* Life Sci. 2007; 81: 1272–9.

Hodgson JM. *Tea flavonoids and cardiovascular disease.* Asia Pac J Clin Nutr. 2008; 17 Suppl 1: 288–90.

Hodnick WF, Kung FS, Roettger WJ, Bohmont CW, Pardini RS. *Inhibition of mitochondrial respiration and production of toxic oxygen radicals by flavonoids. A structure-activity study.* Biochem Pharmacol. 1986; 35: 2345–57.

Hollman PC, Bijsman MN, van Gameren Y, Cnossen EP, de Vries JH, Katan MB.*The sugar moiety is a major determinant of the absorption of dietary flavonoid glycosides in man.* Free Radic Res. 1999; 31: 569–73.

Hong J, Bose M, Ju J, Ryu JH, Chen X, Sang S, Lee MJ, Yang CS. *Modulation of arachidonic acid metabolism by curcumin and related beta-diketone derivatives: effects on cytosolic phospholipase A(2), cyclooxygenases and 5-lipoxygenase.* Carcinogenesis. 2004; 25: 1671–9.

Hou DX, Luo D, Tanigawa S, Hashimoto F, Uto T, Masuzaki S, Fujii M, Sakata Y. *Prodelphinidin B-4 3′-O-gallate, a tea polyphenol, is involved in the inhibition of COX-2 and iNOS via the downregulation of TAK1-NF-kappaB pathway.* Biochem Pharmacol. 2007; 74: 742–51.

Hsieh TC, Juan G, Darzynkiewicz Z, Wu JM. *Resveratrol increases nitric oxide synthase, induces accumulation of p53 and p21 (WAF1/CIP1), and suppresses cultured bovine pulmonary artery endothelial cell proliferation by perturbing progression through S and G2.* Cancer Res. 1999; 59: 479–87.

Hu JP, Calomme M, Lasure A, De Bruyne T, Pieters L, Vlietinck A, Vanden Berghe DA. *Structure-activity relationship of flavonoids with superoxide scavenging activity.* Biol Trace Elem Res. 1995; 47: 327–31.

Huang SM, Wu CH, Yen GC. *Effects of flavonoids on the expression of the pro-inflammatory response in human monocytes induced by ligation of the receptor for AGEs.* Mol Nutr Food Res. 2006; 50: 1129–39.

Huisman A, Van De Wiel A, Rabelink TJ, Van Faassen EE. *Wine polyphenols and ethanol do not significantly scavenge superoxide nor affect endothelial nitric oxide production.* J Nutr Biochem. 2004; 15: 426–32.

Iijima K, Yoshizumi M, Hashimoto M, Akishita M, Kozaki K, Ako J, Watanabe T, Ohike Y, Son B, Yu J, Nakahara K, Ouchi Y. *Red wine polyphenols inhibit vascular smooth muscle cells migration through two distinct signalling pathways.* Circulation. 2002; 105: 2404–10.

Issa AY, Volate SR, Wargovich MJ. *The role of phytochemicals in inhibition of cancer and inflammation: New directions and perspectives.* J Food Compost Anal. 2006; 19: 405–19.

Ivanov V, Ivanova S, Kalinovsky T, Niedzwiecki A, Rath M. *Plant-derived micronutrients suppress monocyte adhesion to cultured human aortic endothelial cell layer by modulating its extracellular matrix composition.* J Cardiovasc Pharmacol. 2008; 52: 55–65.

Jobin C, Bradham CA, Russo MP, Juma B, Narula AS, Brenner DA, Sartor RB. *Curcumin blocks cytokine-mediated NF-kappa B activation and proinflammatory gene expression by inhibiting inhibitory factor I-kappa B kinase activity.* J Immunol. 1999; 163: 3474–83.

Kao YH, Chang HH, Lee MJ, Chen CL. *Tea, obesity, and diabetes.* Mol Nutr Food Res. 2006; 50: 188–210.

Katiyar SK, Afaq F, Azizuddin K, Mukhtar H. *Inhibition of UVB-induced oxidative stress-mediated phosphorylation of mitogen-activated protein kinase signaling pathways in cultured human epidermal keratinocytes by green tea polyphenol (-)-epigallocatechin-3-gallate.* Toxicol Appl Pharmacol. 2001; 176: 110–7.

Kaur C, Kapoor HC. *Anti-oxidants in fruits and vegetables – the millennium's health Int.* J. Food Sci. Technol. 2001; 36: 703–25.

Kawai Y, Nishikawa T, Shiba Y, Saito S, Murota K, Shibata N, Kobayashi M, Kanayama M, Uchida K, Terao J.*Macrophage as a target of quercetin glucuronides in human atherosclerotic arteries: implication in the anti-atherosclerotic mechanism of dietary flavonoids.* J Biol Chem. 2008; 283: 9424–34.

Khanna S, Venojarvi M, Roy S, Sharma N, Trikha P, Bagchi D, Bagchi M, Sen CK. *Dermal wound healing properties of redox-active grape seed proanthocyanidins.* Free Radic Biol Med. 2002; 33: 1089–96.

Kim AR, Cho JY, Zou Y, Choi JS, Chung HY. *Flavonoids differentially modulate nitric oxide production pathways in lipopolysaccharide-activated RAW264.7 cells.* Arch Pharm Res. 2005; 28: 297–304.

Kim HR, Pham HT, Ziboh VA. *Flavonoids differentially inhibit guinea pig epidermal cytosolic phospholipase A2.* Prostaglandins Leukot Essent Fatty Acids. 2001; 65: 281–6.

Kim JH, Shim JS, Lee SK, Kim KW, Rha SY, Chung HC, Kwon HJ. *Microaaray-based analysis of anti-angiogenic activity of demethoxycurcumin on human umbilical vein endothelial cells: crucial involvement of the down-regulation of matrix metalloproteinase.* Jap J Cancer Res. 2002; 93: 1378–85.

Kundu JK, Na HK, Chun KS, Kim YK, Lee SJ, Lee SS, Lee OS, Sim YC, Surh YJ. *Inhibition of phorbol ester-induced COX-2 expression by epigallocatechin gallate in mouse skin and cultured human mammary epithelial cells.* J Nutr. 2003; 133: 3805S–10S.

Kundu JK, Surh YJ. *Epigallocatechin gallate inhibits phorbol ester-induced activation of NF-kappa B and CREB in mouse skin: role of p38 MAPK.* Ann NY Acad Sci. 2007; 1095: 504–12.

Lako J, Wattanapenpaiboon N, Wahlqvist M, Trenerry C. *Phytochemical intakes of the Fijian population.* Asia Pac J Clin Nutr. 2006; 15: 275–85.

Lambert JD, Sang S, Yang CS. *Biotransformation of green tea polyphenols and the biological activities of those metabolites.* Mol Pharm. 2007; 4: 819–25.

Lamy S, Blanchette M, Michaud-Levesque J, Lafleur R, Durocher Y, Moghrabi A, Barrette S, Gingras D, Beliveau R. *Delphinidin, a dietary anthocyanidin, inhibits vascular endothelial rowth factor receptor-2 phosphorylation.* Carcinogenesis. 2006; 27: 989–96.

Lau FC, Shukitt-Hale B, Joseph JA. *Age-related neuronal and behavioral deficits are improved by polyphenol-rich blueberry supplementation*. In: Oxidative Stress and Age-Related Neurodegeneration. Crc Press-Taylor & Francis Group, Boca Raton. 2006; 373–93.

Laughton MJ, Evans PJ, Moroney MA, Hoult JR, Halliwell B. *Inhibition of mammalian 5-lipoxygenase and cyclo-oxygenase by flavonoids and phenolic dietary additives. Relationship to anti-oxidant activity and to iron ion-reducing ability.* Biochem Pharmacol. 1991; 42: 1673–81.

Leese HJ, Semenza G. *On the identity between the small intestinal enzymes phlorizin hydrolase and glycosylceramidase.* J Biol Chem. 1973; 248: 8170–3.

Libby P. *Inflammation in atherosclerosis.* Nature. 2002; 420: 868–74.

Lin YL, Lin JK. *(-)-Epigallocatechin-3-gallate blocks the induction of nitric oxide synthase by down-regulating lipopolysaccharide-induced activity of transcription factor nuclear factor-kappaB.* Mol Pharmacol. 1997; 52: 465–72.

Loke WM, Proudfoot JM, Stewart S, McKinley AJ, Needs PW, Kroon PA, Hodgson JM, Croft KD. *Metabolic transformation has a profound effect on anti-inflammatory activity of flavonoids such as quercetin: lack of association between anti-oxidant and lipoxygenase inhibitory activity.* Biochem Pharmacol. 2008; 75: 1045–53.

Long LH, Clement MV, Halliwell B. *Artifacts in cell culture: rapid generation of hydrogen peroxide on addition of (-)-epigallocatechin, (-)-epigallocatechin gallate, (+)-catechin, and quercetin to commonly used cell culture media.* Biochem Biophys Res Commun. 2000; 273: 50–3.

Loo G. *Redox-sensitive mechanisms of phytochemical-mediated inhibition of cancer cell proliferation.* J Nutr Biochem. 2003; 14: 64–73.

Lorenz M, Wessler S, Follmann E, Michaelis W, Düsterhöft T, Baumann G, Stangl K, Stangl V. *A constituent of green tea, epigallocatechin-3-gallate, activates endothelial nitric oxide synthase by a phosphatidylinositol-3-OH-kinase-, cAMP-dependent protein kinase-, and Akt-dependent pathway and leads to endothelial-dependent vasorelaxation.* J Biol Chem. 2004; 279: 6190–5.

Lu H, Meng X, Yang CS. *Enzymology of methylation of tea catechins and inhibition of catechol-O-methyltransferase by (-)-epigallocatechin gallate.* Drug Metab Dispos. 2003; 31: 572–9.

Luceri C, Caderni G, Sanna A, Dolara P. *Red wine and black tea polyphenols modulate the expression of cycloxygenase-2, inducible nitric oxide synthase and glutathione-related enzymes in azoxymethane-induced f344 rat colon tumors.* J Nutr. 2002; 132: 1376–9.

Lyon CJ, Law RE, Hsueh WA. *Minireview: adiposity, inflammation, and atherogenesis.* Endocrinology. 2003; 144: 2195–200.

Manna SK, Mukhopadhyay A, Aggarwal BB. *Resveratrol suppresses TNF-induced activation of nuclear transcription factors NF-kappa B, activator protein-1, and apoptosis: potential role of reactive oxygen intermediates and lipid peroxidation.* J Immunol. 2000; 164: 6509–19.

Masuda M, Suzui M, Lim JT, Deguchi A, Soh JW, Weinstein IB. *Epigallocatechin-3-gallate decreases VEGF production in head and neck and breast carcinoma cells by inhibiting EGFR-related pathways of signal transduction.* J Exp Ther Oncol. 2002; 2: 350–9.

Miller NJ, Rice-Evans CA. *Anti-oxidant activity of resveratrol in red wine.* Clin Chem. 1995; 41: 1789.

Mink PJ, Scrafford CG, Barraj LM, Harnack L, Hong CP, Nettleton JA, Jacobs DR, Jr. *Flavonoid intake and cardiovascular disease mortality: a prospective study in postmenopausal women.* Am J Clin Nutr. 2007; 85: 895–909.

Miranda PJ, DeFronzo RA, Califf RM, Guyton JR. *Metabolic syndrome: definition,* Mojzis J, Varinska L, Mojzisova G, Kostova I, Mirossay L. *Antiangiogenic effects of flavonoids and chalcones.* Pharmacol Res. 2008; 57: 259–65.

Monteiro R, Calhau C, Silva AO, Pinheiro-Silva S, Guerreiro S, Gärtner F, Azevedo I, Soares R. *Xanthohumol inhibits inflammatory factor production and angiogenesis in breast cancer xenografts.* J Cell Biochem. 2008; 104: 1699–707.

Moon H-S, Lee H-G, Choi Y-J, Kim T-G, Cho C-S. *Proposed mechanisms of (-)-epigallocatechin-3-gallate for anti-obesity.* Chem Biol Interact. 2007; 167: 85–98.

Moreno PR, Purushothaman KR, Fuster V, Echeverri D, Truszczynska H, Sharma SK, Badimon JJ, O'Connor WN. *Plaque neovascularization is increased in ruptured atherosclerotic lesions of human aorta: implications for plaque vulnerability.* Circulation. 2004; 110: 2032–8.
Mursu J, Voutilainen S, Nurmi T, Toumainen T-P, Kurl S, Salonen JT. *Flavonoid intake and the risk of ischaemic stroke and CVD mortality in middle-aged Finnish men: the Kuopio Ischaemic Heart Disease Risk Factor Study.* Br J Nutr. 2008; 1: 1–6.
Negrão MR, Azevedo I, Soares R. *Modulation of Angiogenesis by Beer Polyphenols.* Mol Biol Cell. 2007a; 18: A–1911.
Negrão MR, Íncio J, Lopes R Azevedo I, Soares R. *Evidence for the Effects of Xanthohumol in Disrupting Angiogenic Vessels, but not Stable Ones.* IJBS. 2007b; 3: 279–86.
Oak MH, Chataigneau M, Keravis T, Chataigneau T, Beretz A, Andriantsitohaina R, Stoclet JC, Chang SJ, Schini-Kerth VB. *Red wine polyphenolic compounds inhibit vascular endothelial growth factor expression in vascular smooth muscle cells by preventing the activation of the p38 mitogen-activated protein kinase pathway.* Arterioscler Thromb Vasc Biol. 2003; 23: 1001–7.
Oak MH, El Bedoui J, Schini-Kerth VB. *Antiangiogenic properties of natural polyphenols from red wine and green tea.* J Nutr Biochem. 2005; 16: 1–8.
Ohshima H, Yoshie Y, Auriol S, Gilibert I. *Anti-oxidant and pro-oxidant actions of flavonoids: effects on DNA damage induced by nitric oxide, peroxynitrite and nitroxyl anion.* Free Radic Biol Med. 1998; 25: 1057–65.
Okamoto I, Iwaki K, Koya-Miyata S, Tanimoto T, Kohno K, Ikeda M, Kurimoto M. *The flavonoid Kaempferol suppresses the graft-versus-host reaction by inhibiting type 1 cytokine production and CD8+ T cell engraftment.* Clin Immunol. 2002a; 103: 132–44.
Okamoto T, Yamagishi S, Inagaki Y, Amano S, Koga K, Abe R, Takeuchi M, Ohno S, Yoshimura A, Makita Z. *Angiogenesis induced by advanced glycation end products and its prevention by cerivastatin.* FASEB J. 2002b; 16: 1928–30.
Ollila F, Halling K, Vuorela P, Vuorela H, Slotte JP. *Characterization of flavonoid-biomembrane interactions.* Arch Biochem Biophys. 2002; 399: 103–8.
Opie LH, Lecour S. *The red wine hypothesis: from concepts to protective signalling molecules.* Eur Heart J. 2007; 28: 1683–93.
Ovaskainen ML, Torronen R, Koponen JM, Sinkko H, Hellstrom J, Reinivuo H, Mattila P. *Dietary intake and major food sources of polyphenols in Finnish adults.* J Nutr. 2008; 138: 562–6.
Paquay JB, Haenen GR, Stender G, Wiseman SA, Tijburg LB, Bast A. *Protection against nitric oxide toxicity by tea.* J Agric Food Chem. 2000; 48: 5768–72.
Pérez YY, Jiménez-Ferrer E, Zamilpa A, Hernández-Valencia M, Alarcón-Aguilar FJ, Tortoriello J, Román-Ramos R. *Effect of a polyphenol-rich extract from Aloe vera gel on experimentally induced insulin resistance in mice.* Am J Chin Med. 2007; 35: 1037–46
Rahman I, Biswas SK, Kirkham PA. *Regulation of inflammation and redox signaling by dietary polyphenols.* Biochem Pharmacol. 2006; 72: 1439–52.
Ratty AK, Das NP.*Effects of flavonoids on nonenzymatic lipid peroxidation: structure-activity relationship.* Biochem Med Metab Biol. 1988; 39: 69–79.
Rice-Evans C. *Flavonoid anti-oxidants.* Curr Med Chem. 2001; 8: 797–807.
Rice-Evans CA, Miller NJ, Paganga G. *Structure-anti-oxidant activity relationships of flavonoids and phenolic acids.* Free Radic Biol Med. 1996; 20:933.
Rossi M, Negri E, Lagiou P, Talamini R, Dal Maso L, Montella M, Franceschi S, La Vecchia C. *Flavonoids and ovarian cancer risk: A case-control study in Italy.* Int J Cancer. 2008; 123: 895–8.
Rossi M, Negri E, Talamini R, Bosetti C, Parpinel M, Gnagnarella P, Franceschi S, Dal Maso L, Montella M, Giacosa A, La Vecchia C. *Flavonoids and colorectal cancer in Italy.* Cancer Epidemiol Biomarkers Prev. 2006; 15: 1555–8.
Ruiz PA, Braune A, Hölzlwimmer G, Quintanilla-Fend L, Haller D.*Quercetin inhibits TNF-induced NF-kappaB transcription factor recruitment to proinflammatory gene promoters in murine intestinal epithelial cells.* J Nutr. 2007; 137: 1208–15.

Santangelo C, Varì R, Scazzocchio B, Di Benedetto R, Filesi C, Masella R. *Polyphenols, intracellular signalling and inflammation.* Ann Ist Super Sanita. 2007; 43: 394–405.

Sato M, Miyazaki T, Kambe F, Maeda K, Seo H. *Quercetin, a bioflavonoid, inhibits the induction of interleukin 8 and monocyte chemoattractant protein-1 expression by tumor necrosis factor-alpha in cultured human synovial cells.* J Rheumatol. 1997; 24: 1680–4.

Scalbert A, Williamson G. *Dietary intake and bioavailability of polyphenols.* J Nutr. 2000; 130: 2073S–85S.

Scholz S, Williamson G. *Interactions affecting the bioavailability of dietary polyphenols in vivo.* Int J Vitam Nutr Res. 2007; 77: 224–35.

Shetty AK, Rashmi R, Rajan MGR, Sambaiah K, Salimath PV. *Antidiabetic influence of quercetin in streptozotocin-induced diabetic rats.* Nutr Res. 2004; 24: 373–81.

Shimoi K, Okada H, Furugori M, Goda T, Takase S, Suzuki M, Hara Y, Yamamoto H, Kinae N. *Intestinal absorption of luteolin and luteolin 7-O-beta-glucoside in rats and humans.* FEBS Lett. 1998; 438: 220–4.

Shirai M, Moon JH, Tsushida T, Terao J. *Inhibitory effect of a quercetin metabolite, quercetin 3-O-beta-D-glucuronide, on lipid peroxidation in liposomal membranes.* J Agric Food Chem. 2001; 49: 5602–8.

Singh M, Arseneault M, Sanderson T, Murthy V, Ramassamy C. *Challenges for research on polyphenols from foods in Alzheimer's disease: bioavailability, metabolism, and cellular and molecular mechanisms.* J Agric Food Chem. 2008; 56: 4855–73.

Singh S, Khar A. *Biological effects of curcumin and its role in cancer chemoprevention and therapy.* Anticancer Agents Med Chem. 2006; 6:259–70.

Soares R, Azevedo I. *Inhibition of S1P by polyphenols prevents inflammation and angiogenesis: NFkappaB, a downstream effector?* Free Radic Biol Med. 2007; 42: 311.

Soares R, Costa C. *Angiogenesis and inflammatory diseases: current concepts and therapeutic perspectives.* In: Maragoudakis ME; Papadimitriou E (ed.) Angiogenesis. Basic science and clinical applications, 1st edn. Transworld Research Network. 2007; 511–47.

Soares S, Mateus N, De Freitas V. *Interaction of different polyphenols with bovine serum albumin (BSA) and human salivary alpha-amylase (HSA) by fluorescence quenching.* J Agric Food Chem. 2007; 55: 6726–35.

Song Y, Manson JE, Buring JE, Sesso HD, Liu S. *Association of dietary flavonoids with risk of type 2 diabetes, and markers of insulin resistence and systemic inflammation in women: a prospective study and cross-sectional analysis.* J Am Coll Nutr. 2005; 24: 376–84.

Soobrattee MA, Neergheen VS, Luximon-Ramma A, Aruoma OI, Bahorun T. *Phenolics as potential antioxidant therapeutic agents: mechanism and actions.* Mutat Res. 2005; 579: 200–13.

Spencer JP, Schroeter H, Crossthwaithe AJ, Kuhnle G, Williams RJ, Rice-Evans C. *Contrasting influences of glucuronidation and O-methylation of epicatechin on hydrogen peroxide-induced cell death in neurons and fibroblasts.* Free Radic Biol Med. 2001a; 31: 1139–46.

Spencer JP, Schroeter H, Kuhnle G, Srai SK, Tyrrell RM, Hahn U, Rice-Evans C. *Epicatechin and its in vivo metabolite, 3'-O-methyl epicatechin, protect human fibroblasts from oxidative-stress-induced cell death involving caspase-3 activation.* Biochem J. 2001b; 354: 493–500.

Stangl V, Dreger H, Stangl K, Lorenz M. *Molecular targets of tea polyphenols in the cardiovascular system.* Cardiovasc Res. 2007; 73: 348–58.

Steffen Y, Gruber C, Schewe T, Sies H. *Mono-O-methylated flavanols and other flavonoids as inhibitors of endothelial NADPH oxidase.* Arch Biochem Biophys. 2008; 469: 209–19.

Stoclet JC, Chataigneau T, Ndiaye M, Oak MH, El Bedoui J, Chataigneau M, Schini-Kerth VB. *Vascular protection by dietary polyphenols.* Eur J Pharmacol. 2004; 500: 299–313.

Surh YJ, Chun KS, Cha HH, Han SS, Keum YS, Park KK, Lee SS. *Molecular mechanisms underlying chemopreventive activities of anti-inflammatory phytochemicals: down-regulation of COX-2 and iNOS through suppression of NF-kappa B activation.* Mutat Res. 2001; 480–481: 243–68.

Surh YJ. *Cancer chemoprevention with dietary phytochemicals.* Nat Rev Cancer. 2003; 3: 768–80.

Tabernero M, Serrano J, Saura-Calixto F. *Dietary fiber intake in two European diets with high (copenhagen, Denmark) and low (Murcia, Spain) colorectal cancer incidence.* J Agric Food Chem. 2007; 55: 9443–9.

Terao J, Yamaguchi S, Shirai M, Miyoshi M, Moon JH, Oshima S, Inakuma T, Tsushida T, Kato Y. *Protection by quercetin and quercetin 3-O-beta-D-glucuronide of peroxynitrite-induced antioxidant consumption in human plasma low-density lipoprotein.* Free Radic Res. 2001; 35: 925–31.

Terra X, Valls J, Vitrac X, Mérrillon JM, Arola L, Ardèvol A, Bladé C, Fernandez-Larrea J, Pujadas G, Salvadó J, Blay M. *Grape-seed procyanidins act as antiinflammatory agents in endotoxin-stimulated RAW 264.7 macrophages by inhibiting NFkB signaling pathway.* J Agric Food Chem. 2007; 55: 4357–65.

Tsuda T.*Regulation of adipocyte function by anthocyanidins; possibility of preventing the metabolic syndrome.* J Agric Food Chem. 2008; 56: 642–6.

van Acker SA, de Groot MJ, van den Berg DJ, Tromp MN, Donne-Op den Kelder G, van der Vijgh WJ, Bast A. *A quantum chemical explanation of the anti-oxidant activity of flavonoids.* Chem Res Toxicol. 1996; 9: 1305–12.

Vanden Berghe W, Ndlovu MN, Hoya-Arias R, Dijsselbloem N, Gerlo S, Haegeman G. *Keeping up NF-kappaB appearances: epigenetic control of immunity or inflammation-triggered epigenetics.* Biochem Pharmacol. 2006;72:1114–31.

Van Duyn MA, Pivonka E. *Overview of the health benefits of fruit and vegetable consumption for the dietetics professional: selected literature.* J Am Diet Assoc. 2000; 100: 1511–21.

Vennat B, Bos MA, Pourrat A, Bastide P. *Procyanidins from tormentil: fractionation and study of the anti-radical activity towards superoxide anion.* Biol Pharm Bull. 1994; 17: 1613–5.

Vinson JA, Teufel K, Wu N. *Red wine, dealcoholizes red wine, and especialy grape juice, inhibit atherosclerosis in a hamster model.* Atherosclerosis. 2001; 156: 67–72.

Walle T. *Methylation of dietary flavones greatly improves their hepatic metabolic stability and intestinal absorption.* Mol Pharm. 2007; 4: 826–32.

Weisberg SP, Leibel R, Tortoriello DV. *Dietary curcumin significantly improves obesity-associated inflammation and diabetes in mouse models of diabesity.* Endocrinology. 2008; 149: 3549–58.

Weisberg SP, McCann D, Desai M, Rosenbaum M, Leibel RL, Ferrante AW Jr. *Obesity is associated with macrophage accumulation in adipose tissue.* J Clin Invest. 2003; 112: 1796–808.

Wheeler DS, Catravas JD, Odoms K, Denenberg A, Malhotra V, Wong HR. *Epigallocatechin-3-gallate, a green tea-derived polyphenol, inhibits IL-1 beta-dependent proinflammatory signal transduction in cultured respiratory epithelial cells.* J Nutr. 2004; 134: 1039–44.

Wilkinson AP, Gee JM, Dupont MS, Needs PW, Mellon FA, Williamson G, Johnson IT. *Hydrolysis by lactase phlorizin hydrolase is the first step in the uptake of daidzein glucosides by rat small intestine in vitro.* Xenobiotica. 2003; 33: 255–64.

Wilson T, Knight TJ, Beitz DC, Lewis DS, Engen RL. *Resveratrol promotes atherosclerosis in hypercholesterolemic rabbits.* Life Sci. 1996; 59: PL15–21.

Wu L-Y, Juan C-C, Ho L-T, Hsu Y-P, Hwang LS. *Effect of green tea supplementation on insulin sensitivity in Sprague-Dawley Rats.* J Agric Food Chem. 2004; 52: 643–8.

Wu X, Pittman HE, 3rd, McKay S, Prior RL. *Aglycones and sugar moieties alter anthocyanin absorption and metabolism after berry consumption in weanling pigs.* J Nutr. 2005; 135: 2417–24.

Xagorari A, Roussos C, Papapetropoulos A. *Inhibition of LPS-stimulated pathways in macrophages by the flavonoid luteolin.* Br J Pharmacol. 2002; 136: 1058–64.

Yoon JH, Baek SJ. *Molecular targets of dietary polyphenols with anti-inflammatory properties.* Yonsei Med J. 2005; 46: 585–96.

Yu R, Hebbar V, Kim DW, Mandlekar S, Pezzuto JM, Kong AN. *Resveratrol inhibits phorbol ester and UV-induced activator protein 1 activation by interfering with mitogen-activated protein kinase pathways.* Mol Pharmacol. 2001; 60: 217–24.

Zhang Q, Tang X, Lu Q, Zhang Z, Rao J, Le AD. *Green tea extract and (-)-epigallocatechin-3-gallate inhibit hypoxia- and serum-induced HIF-1α protein accumulation and VEGF expression in human cervical carcinoma and hepatoma cels.* Mol Cancer Ther. 2006; 5: 1227–38.

Zhou B, Miao Q, Yang L, Liu ZL. *Antioxidative effects of flavonols and their glycosides against the free-radical-induced peroxidation of linoleic acid in solution and in micelles.* Chemistry. 2005; 11: 680–91.

Chapter 9
Metabolic Syndrome: Practical Implications of a Concept

Cassiano Abreu-Lima

Contents

9.1 Epidemiologic Transition: The Driving Force of Metabolic Syndrome

Following the reduction of cardiovascular disease mortality observed in high-income regions of the world in the last decades of the 20th century, humanity is facing the threat of a global epidemics of obesity and diabetes that will entail a new increase in atherosclerotic vascular disease morbidity and mortality.

This phenomenon is well explained by the theory of Omran in 1971 and later developed by Olshansky and Ault, relating demographic, economic and social evolution to morbidity and mortality (Omran 1971; Olshansky and Ault 1986). The concept, masterly divulged by Gaziano (2008) in atlas and textbooks, considers four phases in this evolution and takes the United States of America (USA) as reference standard for demonstration, believing that most of what could be observed can be broadening generalized to other countries that followed approximately the same pattern of development, within a more or less marked time-lag.

The 1st phase is appropriately named *the age of pestilence and famine*. Its main characteristics are an agrarian economy and a predominant rural style of living, a low life expectancy at birth and a high infant and child mortality, being

C. Abreu-Lima (✉)
Department of Cardiology, Faculty of Medicine, University of Porto,
Al. Prof. Hernâni Monteiro, 4200-319 Porto, Portugal
e-mail: cassal@netcabo.pt

R. Soares, C. Costa (eds.), *Oxidative Stress, Inflammation and Angiogenesis in the Metabolic Syndrome,* DOI 10.1007/978-1-4020-9701-0_9,

undernutrition and infectious diseases (tuberculosis, pneumonia and diarrhoeal disease) the main causes of death. Cardiovascular disease and cancer account for less than 10% of all deaths and rheumatic heart disease dominates the cardiovascular morbidity scenery.

To this time period - through which in fact humanity evolved along most of the recorded history and that elapsed in the USA during the 19th century – followed the so-called *age of receding pandemics* (1900–1930 in the USA), when a progressive decrease in malnutrition and infectious disease death rates, brought about by income growth and public health and sanitization measures, occurred. A significant decline in infant and child mortality was one of the most impressive signs of this progress. By then, cardiovascular disease changed both under the point of view of morbidity and causes of death: to rheumatic heart disease, hypertension, coronary heart disease and stroke were added as prominent cardiovascular disease expressions, and mortality of cardiovascular origin began to build-up (10–35%).

Expansion of industrialization, urbanization and mechanization, economic growth and increasing wealth that followed (USA 1930–1965) prompted a major change in the style of living that paved the way to the following phase: *the age of degenerative and man-made diseases*. Its main characteristics are decreased rates of physical activity and high caloric intake from saturated fat and sugar. Overweight and obesity and cigarette smoking increase. This dreadful combination favoured the development of dyslipidaemia and glucometabolic disturbances - impaired fasting glucose, glucose tolerance and glucose homeostasis, and full blown diabetes - and set the stage for the entrance in scene of metabolic syndrome, a highly atherogenic milieu that continued to expand to our days and afar. The predominant cardiovascular diseases are now the two main clinical expressions of atherosclerosis, coronary heart disease and stroke. Mortality from chronic, non-communicable diseases exceeds mortality from malnutrition and infectious diseases and life expectancy increases beyond 50 years. Cancer morbidity and mortality also increased. Cardiovascular disease mortality attains its highest level: 35–65%.

The following and fourth phase is called *the age of delayed degenerative diseases* and took place in the USA and other high income countries during the last 35 years of the 20th century. There and then, important developments in patient care took place, thanks to notable technological advances for the treatment of acute cardiovascular ailments and to preventive efforts aiming at reducing cigarette smoking and improving better control of high blood pressure and high cholesterol levels, mainly through very efficient pharmacologic treatment strategies. Life expectancy continued to grow. Cancer mortality increased. Due to increased survival to acute cardiac assaults and increased life span, an outbreak of heart failure, whose costs of treatment rapidly led the list of health expenses, began to emerge in these countries. Cardiovascular mortality decreased progressively. However, owing to the time lag in development between high-income and low and middle income-countries and the corresponding later shift in these of the communicable to non-communicable diseases' rates ratio, cardiovascular mortality is growing steadily in developing regions, and this has a major impact in global cardiovascular mortality, considering that 85% of the world population lives there. Besides, even in high-income countries, deceleration of cardiovascular mortality reduction, that in fact is to be ascribed more to advances in medical care than to widespread adoption of healthy lifestyles, began to be noticed in the last years of the past century and in the present decade. The reason for this slowing decline of cardiovascular mortality, may well reside in stagnation or even increment of prevalence of cardiovascular risk factors, namely smoking, physical inactivity, overweight and obesity and glucometabolic disturbances (Gaziano 2008), and metabolic syndrome (MS). Cumulative evidence suggests that the number of people engaged in leisure time physical activity keeps increasing on and on. This is somehow true in absolute terms. However, when compared to the more rapidly growing number of inactive people, brought about by worldwide increase in urbanization, industrialization and mechanization, the proportion of physical activists is in fact less and less, reasonably allowing to foresee that in the near future, only a scarce minority of the world population will meet the minimal physical activity recommendations (Gaziano 2008), a behaviour that along with the alarming pace at which overweight and obesity are increasing worldwide will result in increases in glucometabolic and cardiovascular disorders.

9.2 The Future: The Threat of Global Epidemics of Physical Inactivity, Obesity and Diabetes

What comes next? Are we in the dawn of an age whose main characteristics are very high prevalence rates of population inactivity, obesity and diabetes leading to new and alarming increases in age-adjusted cardiovascular mortality (Gaziano 2008)? Fears are not without reason. Obesity and overweight prevalence that already approached 45% of the American population in the beginning of the 1960's, continued to grow to exceed 60% in 1999 (Cefalu 2008). The rate of this growth also increased, denoting not a linear, but an exponential pattern of expansion. More recently, prevalence in obesity and diabetes among US adults was estimated to increase 5.6% and 8.2%, respectively, in a single year, between 2000 and 2001 (Mokdad et al. 2003). In this same nationwide survey, the prevalence of body mass index of 40 or higher in 2001 was 2.3% and overweight and obesity were shown to be significantly associated with diabetes, hypertension, hypercholesterolaemia, asthma, arthritis and poor health status (Mokdad et al. 2003), indicating that its prevention and effective treatment can be rewarding in multiple ways. It seems that the increase in the number of diabetics in the last decades largely due to the increase in obesity does not look to have spared any world region. Projections suggest that in 2030 the number of people with diabetes may attain 366 million (Wild et al. 2004). In Portugal, a survey between 2003 and 2005 that included 8116 participants aged 18–64 showed that 39.4% were overweight and 14% obese; roughly 46% of the sample had high waist circumference. Compared to a previous survey by the same group (1995–1998), overweight/obesity prevalence increased by about 8% (do Carmo et al. 2008). In a nationwide survey, Cortez-Dias and collaborators estimated that the age and sex-adjusted prevalence of MS (ATP-III criteria) in Portugal hits 1/3 of the adult population (29.4%: 31.2% in women and 27.5% in men) (Cortez-Dias et al. 2007). Interestingly enough, these authors found that participants in the study had an increasingly high prevalence of waist circumference, hypertriglycerides, low HDL-cholesterol, high blood pressure, fasting hyperglycaemia and MS along three strata of the normal weight range: 18.5–20.9; 21–22.9; 23–24.9 kg/m^2(Martins et al. 2007), stressing the existence of a continuum of cardiometabolic risk from low-normal weight to obesity.

A troubling increasing prevalence of inactivity, obesity, hypertension and diabetes is observed also in children and adolescents and young adults, in developed as well as in developing countries, a phenomenon calling for special attention to these age groups in the design of preventive strategies at the individual as well as population level. The number of European Union schoolchildren who are overweight rises by about 400,000 per year (International Obesity Task Force, 2006). In Portugal, Padez et al. (2004) estimated the prevalence of overweight and obesity among 4,511 male and female children from 7 to 9 years to be 31.6% (also greater in females as among adults) and disclosed significant associations between the biotype and hours spent regarding television and playing computer games (direct association), birth weight (direct), parents obesity (direct), parents education level (indirect) and number family children (indirect). The authors concluded that intervention strategies to fight against overweight and obesity should involve the family as a whole. Also in Portugal, countrywide observation of registries of weight, height and body-mass index of young recruits showed increases of 4.0%, 0.4% and 2.6%, respectively, between 1994–1995 and 1998–1999. In these four years, overweight and obesity increased from 15% to 22%, and obesity alone, from 1.4% to 2.3% (Nobre et al. 2004).

9.3 Metabolic Syndrome: Impact of a Concept and Clinical Usefulness

Being an appealing concept, metabolic syndrome, an easy to suspect and to recognize cardiovascular risk factors' aggregation, strongly stimulated research and had a positive impact on clinical practice: throughout an enormous number of continued medical education actions, the concept has been widely divulged among clinicians and contributed to arouse their awareness of cardiovascular risk and risk factors control that certainly changed and improved their practice.

However, MS should be viewed more as pathophysiological entity than as a tool to calculate cardiovascular risk with reasonable precision, which it is not. Patients' management varies with their risk level and this is the reason why we need such tools to assess cardiovascular risk in the individual patient.

Framingham Heart Study provided such tools. Their basis is an equation developed through a powerful multivariate procedure, known as multiple logistic regression analysis, which relates risk factors to the probability of a dichotomous outcome such as the occurrence or not of disease or death. Most of the risk calculators at the disposal of clinicians are built-up upon such equations, namely those developed by the European Society of Cardiology: *Score* that uses coloured tables and *Heart Score*, a computer program (European Society of Cardiology). These European tools were designed to estimate the ten year risk (probability) of fatal cardiovascular disease in individuals without overt cardiovascular disease or diabetes, according to the region of Europe the patient belongs to (low risk region tables for Portugal) and patient's sex, age, smoking status and usual systolic blood pressure and total cholesterol levels.

Clinical risk profiles defined by dichotomized variables such us the presence or absence of modifiable conditions like obesity, physical inactivity, hypertension, diabetes, left ventricular hypertrophy, atrial fibrillation and smoking status are useful for rough risk estimation and to motivate clinicians and patients to undertake cardiovascular risk reduction measures. Besides, they also help to identify targets in intervention strategies at a population level, aiming at reducing the burden of atherosclerotic cardiovascular morbidity and mortality. MS is one such profile. However, owing to the fact that vascular risk is a continuum across any continuous variable risk factor, the transformation of these continuous variables in discrete, dichotomous ones by use of threshold values, considerably weakens their predictive power. So, when it comes to risk stratification on more solid grounds, the use of risk calculators based on Framingham equations is recommended as stated before. In fact, MS criteria do not include three major factors for risk cardiovascular assessment: age, smoking status and actual level of LDL cholesterol. As a consequence, patients with MS by any set of criteria can be at any level of risk (Preiss and Sattar 2007), as evaluated by powerful, validated risk calculation tools based on Framingham equation.

Anyhow, MS is an invaluable concept. It is a clinical vignette that instantly, through a single look, awakes the need for a risk reduction attitude and easily shows the main targets of it.

The other and more meaningful face of the coin shows a regrettable and dreadful perspective: MS is the cameo of man-made diseases. It is the ambulant icon of human biological mal-adaptation to unhealthy life-styles brought about by the so called civilization which in scarce decades put in risk million years of perfecting biological evolution.

References

Omran AR. *The epidemiologic transition. A theory of the epidemiology of the population change.* Milbank Mem Fund Q. 1971; 49: 509–38.

Olshansky SJ, Ault AB.*The fourth stage of the epidemiologic transition: the age of delayed degenerative diseases.* Milbank Q. 1986; 64: 355–91.

Gaziano JM. *Global Burden of Cardiovascular Disease.* In Libby P, Bonow RO, Mann DL, Zipes DP (ed.) Braunwald's Heart Disease, Part I, Chapter 1 pp. 1–22, eighth edition, Saunders – Elsevier, 2008.

Cefalu WT. *Classificação e evolução dos estados de risco cardiometabólico aumentado.* In Cefalu WT, Cannon CP. Atlas de Risco Cardiometabólico. Portuguese Edition. EUROMEDICE, Edições Médicas Lda, Algés, 2008.

Cortez-Dias N, Martins S, Fiuza M. *Metabolic syndrome: an evolving concept.* Rev Port Cardiol. 2007; 26: 1409–21.
Mokdad AH, Ford ES, Bowman BA et al. *Prevalence of obesity, diabetes, and obesity-related health risk factors.* JAMA. 2003; 289: 76–9.
Wild S, Roglic G, Green A, Sicree R, King H. *Global prevalence of diabetes. Estimates for the year 2000 and projections for 2030.* Diabetes Care. 2004; 27: 1047–53.
do Carmo I, dos Santos O, Camolas J et al. Obes Rev. 2008; 9: 11–9.
http://www.spc.pt/congressoXXVIII/documentos/jornalabril_3.pdf
Martins S, Cortez-Dias N, Fiúza M. *On behalf of the Valsim Investigators.* Circulation. 2007; 116: II 804 (abstract 3554).
International Obesity Task Force. http://www.iotf.org/index.asp.
Padez C, Mourão I, Moreira P, Fernnades T, Marques V. *Obesidade infantil em Portugal.* http://www.adexo.pt/pdf/Obesidade%20infantil.pps#309.
Nobre EL, Jorge Z, Macedo A, de Castro JJ. *Tendências do peso em Portugal no final do século XX. Estudo de coorte de jovens do sexo masculino.* Acta Med Port. 2004; 17: 205–9.
European Society of Cardiology. http://www.escardio.org/policy/prevention/tools/health-toolkit/pages/heartscore.aspx.
Preiss D, Sattar N. *Metabolic syndrome, dysglicaemia and cardiovascular disease: making sense of the evidence.* Heart 2007; 93: 1493–96.

List of Abbreviations

2-NBDG	glucose analog
AA	arachidonic acid
ACE	angiotensin converting enzyme
ADP	adenosine diphosphate
AGE	advanced glycation end-products
AMP	adenosine monophosphate
Ang-2	angiopoietin-2
AP-1	activator protein-1
Apo	apolipoprotein
ARB	angiotensin receptor blockers
AT	adipose tissue
ATF-6	activating transcription factor-6
ATP	adenosine triphosphate
ATP III	adult treatment panel III
BBB	blood-brain barrier
bFGF	basic fibroblast growth factor
BM	bone marrow
BMI	body mass index
BP	blood pressure
BRB	blood-retinal barrier
BUI	brain uptake index
CAD	coronary artery disease
CAM	chick embryo choroallantoic membrane
cAMP	cyclic adenosine monophosphate
CARDIA	coronary artery risk development in young adults
CETP	cholesteryl ester transfer protein
cGMP	cyclic guanosine monophosphate
CHD	coronary heart disease
CoA	coenzyme A
COMT	catechol-O-methyltransferase
COX	cyclooxygenase
CREB	cAMP response element binding protein
CRP	C-reactive protein

CVD	cardiovascualr disease
DNA	deoxyribonucleic acid
EC	endothelial cells
ECM	extracellular matrix
ED	endothelial dysfunction
EGCG	epigallocatechin-3-gallate
EGF	epidermal growth factor
EGIR	European Group of Insulin Resistance
eNOS	endothelial NOS
EOCs	early outgrowth colonies
EPCs	endothelial progenitor cells
ER	endoplasmic reticulum
ERK	extracellular signal-regulated protein kinase
ETC	electron transport chain
FFAs	free fatty acids
FGF	fibroblast growth factor
GLUT	glucose transporter
GPX	glutathione peroxidase
GSH	glutathione
GSSG	oxidized glutathione
H_2O_2	hydrogen peroxide
HAECs	human aortic endothelial cells
HASMC	human aortic vascular smooth muscle cells
HAVECs	human autologous venous endothelial cells
HCAECs	human coronary artery endothelial cells
HDL	high density lipoprotein
HETE	hydroxyeicosatetraenoic acid
HOXA9	homeobox A9
HPA	hypothalamus-pituitary-adrenal cortex
HRECs	human retinal endothelial cells
HSL	hormone sensitive lipase
HUVEC	human umbilical vein endothelial cells
IκB	inhibitor κB proteins
I-CAM	intercellular adhesion molecule
IFNγ	interferon γ
IGF-1	insulin-like growth factor
IKK	inhibitor of nuclear factor κB kinase
IL	interleukin
iNOS	inducible NOS
IR	insulin resistance
IRAK	IL-1β-mediated IL-1β receptor-associated kinase
IRE-1	inositol-requiring enzyme-1
IRS	insulin receptor substrate
IXN	isoxanthohumol
JNK	c-Jun amino terminal kinase

LDL	low density lipoprotein
LOCs	late outgrowth colonies
LOX	lipo-oxygenases
LPS	lypopolisaccharide
LT	leukotriene
MAO	monoaminoxidase
MAPK	mitogen-activated protein kinase
MAPKK	MAPK kinase
MAPKKK	MAPKK kinase
MCP-1	monocyte-chemoattractantprotein-1
MIP1α	macrophage inflammatory molecule 1α
MMP	matrix metalloproteinase
MPO	myeloperoxidase
mRNA	messenger ribonucleic acid
MS	metabolic syndrome
MT1-MMP	membrane type 1-MMP
mTOR	mammalian target of rapamycin
NADPH	nicotinamide adenine dinucleotide phosphate
NAG-1	NSAID-activated gene-1
NECP	national cholesterol education program
NEFA	non-esterified fatty acid
NFκB	nuclear factor kappaB
nNOS	neuronal NOS
NO	nitric oxide
NOS	nitric oxide synthase
Nox	NAD(P)H oxidases
NSAID	nonsteroidal anti-inflammatory drug
ox-LDL	oxidized-LDL
PAI-1	plasminogen activator inhibitor type-1
PDGF	platelet-derived growth factor
PEDF	pigment epithelium-derived factor
PERK	PKR-like eukaryotic initiation factor 2a kinase
PG	prostaglandins
PGE_2	prostaglandin E_2
PGI_2	prostacyclin
PI3K	phosphoinositide-3 kinase
PKA	protein kinase A
PKB	protein kinase B
PKC	protein kinase C
PLA_2	phospholipase A_2
PlGF	placental growth factor
PNMT	phenylethanolamine-N-methyl transferase
PON-1	paraoxonase-1
PPARγ	peroxisome proliferator-activated receptor γ
PUFA	polyunsaturated fatty acids

RAGE	receptor of advanced glycation end-products
RNS	reactive nitrogen species
ROS	reactive oxygen species
S1P	platelet-derived lipid sphingosine-1-phosphate
SDF-1	stromal derived factor-1
sdHDL	small dense HDL
sdLDL	small dense LDL
sGC	soluble guanylate ciclase
SGLT	glucose co-transporters
SIRT	sirtuin
SMC	smooth muscle cells
SOD	superoxide dismutase
SREBP	sterol regulatory element binding protein
SRH	spontaneously hypertensive rats
STAT-1	signal transducer and activator of transcription 1
T2DM	type 2 diabetes mellitus
TGFβ	transforming growth factor β
TLR	toll-like receptor
TNF	tumour necrosis factor
tPA	tissue type of plasminogen activator
TPL	4-hydroxy tempol
Tsp	thrombospondin
TXA_2	thromboxan A_2
UCP	uncoupling proteins
uPA	urokinase type of plasminogen activator
UPR	unfolded protein response
V-CAM	vascular adhesion molecule
VEC	vascular endothelial cell
VEGF	vascular endothelial growth factor
VEGFR	vascular endothelial growth factor receptor
VLDL	very low density lipoprotein
VLDLR	receptor of very low density lipoproteins
VSMC	vascular smooth muscle cells
WHO	World Health Organization
XN	xanthohumol

Glossary

Activator protein-1 Heterodimeric protein that functions as a transcription factor upregulating the transcription of genes containing the 12-O-tetradecanoylphorbol-13-acetate response element

Acute inflammation Short-term process characterized by the classic signs of inflammation – swelling, redness, pain, heat, and loss of function – due to the infiltration of the tissues by plasma and leukocytes

Adenosine Multifunctional endogenous nucleoside that modulates many biochemical processes, such as energy transfer – as adenosine triphosphate (ATP) and adenosine diphosphate (ADP) – as well as in signal transduction as cyclic adenosine monophosphate, cAMP. It is also described as an inhibitory neurotransmitter

Adipocyte Primary cell type that composes the adipose tissue, specialized in storing lipids and produces specific secretions

Adipogenesis Process of differentiation of adipocyte precursor cells, or preadipocytes, into mature adipocytes

Adipokine Cytokine secreted by the adipose tissue

Adiponectin Protein hormone secreted mainly by the adipose tissue, involded in metabolic regulation of carbohydrate and lipid metabolism; it is decreased in obese states

Adrenaline Also known as epinephrine; a hormone mainly secreted by adrenal medulla, involved in metabolic and cardiovascular homeostasis

Adult vasculogenesis Postnatal adaptation of the process of embryonic vasculogenesis, in which circulating Endothelial Progenitor Cells are mobilized from the bone marrow to the peripheral circulation and to neovascular sites, where they participate in the development of vascular networks by differentiating into mature endothelial cells

Advanced glycation end-products Molecules modified from the spontaneous reaction with carbohydrates commonly found in tissues of diabetics

Aldosterone Hormone secreted by the adrenal cortex which affects blood pressure and saline balance

Adrenoceptor Class of G protein-coupled receptors through which noradrenaline and adrenaline act as important neurotransmitters and hormones in the central nervous system and in the periphery

Alpha 2-adrenoceptor Subtype of alpha adrenoceptor

Allostasis The ongoing adaptive efforts of the body to maintain homeostasis in response to stressors

AMP – activated protein kinase Enzyme expressed in a number of tissues that plays a role in cellular energy homeostasis

Android obesity Also know as "apple-shaped obesity"; obesity of the male type that shows a dominant visceral and upper thoracic distribution of adipose tissue

Angiogenesis Complex multistep process that enables the formation of new blood vessels from pre-existing ones

Angiopoietin Protein that plays a role in angiogenesis and vascular development. Angiopoietins bind to specific receptors at the endothelial cell membrane mediating interaction between endothelium and extracellular environment

Angiotensin Family of peptides that act as potent direct vasoconstrictors to narrow blood vessels; stimulates the release of aldosterone from the adrenal cortex

Angiotensin converting enzyme inhibitors Drugs that inhibit the production of angiotensin II and lower blood pressure

Angiotension receptor blockers Medications that block the action of angiotensin II, resulting in blood vessels dilation and reduction in blood pressure

Antioxidants Natural or synthetic substances that prevent or delay any processes of oxidation that can be damaging to cells and tissues

Apoptosis Programmed cell death; occurs when a cell is damaged beyond repair. Involves a series of biochemical events leading to a variety of cellular morphological changes, characterized by cleavage of chromosomal DNA, chromatin condensation, and fragmentation of both the nucleus and the cell

Atherosclerosis Chronic inflammatory condition in which arteries undergo gradual intima thickness, causing decreasing elasticity, narrowing, and reduced blood supply

Bax Pro-apoptotic protein of the Bcl-2 gene family; key component for cellular apoptosis through mitochondrial stress. Increases membrane permeability, leading to the release of cytochrome c from mitochondria, activation of caspase-9 and initiation of the caspase activation pathway for apoptosis

Beta 2-adrenoceptors Subtype of beta adrenoceptor

11 beta-hydroxy-steroid dehydrogenase Enzyme that catalyzes the interconversion of active glucocorticoids (such as cortisol and corticosterone) and their inactive forms (such as cortisone and 11-dehydrocorticosterone)

Bioavailability Measurement of the extent of certain substance that reaches the systemic circulation and is available at the target site

Bone marrow The soft, living tissue that fills most bone cavities and contains hematopoietic stem cells, from which all red and white blood cells evolve, and mesenchymal stem cells

Catecholamines Amines derived from tyrosine that act as neurotransmitters or hormones; the most important are adrenaline, noradrenaline and dopamine

cAMP response element-binding protein Transcription factor that binds to DNA sequences called cAMP response elements to increase or decrease the transcription of certain genes

Catechol-*o*-methyltransferase Key enzyme in the degradation of catechols, such as catecholamines, by transfering a methyl group from adenosylmethionine

CD133 Also known as AC133; 97 kDa cell surface glycoprotein with 5 transmembrane domains, expressed by hematopoietic, neural and embryonic stem cells and hematopoietic and endothelial progenitor cells. It was also found in several tumours, including leukemias and brain tumours

CD34 Cluster of differentiation; 120kDa transmembrane glycoprotein cell surface present in hematopoietic stem and progenitor cells, Endothelial Progenitor Cells and mature endothelial cells. Potential adhesion molecule with a role in early hematopoiesis by mediating the attachment of stem cells to the bone marrow extracellular matrix or directly to stromal cells

Cell senescence The limited capacity of cells to divide beyond a finite number of population doublings (finite growth potential). Irreversible growth-arrest state that depends on the age or cell doublings of a cell

Cell therapy Describes the process of introducing new cells into a tissue in order to treat a specific pathology; there are several potential types of cell therapy: using autologous (from the patient) or allogeneic (from another donor) hematopoietic stem or progenitor cells; mesenchymal stem cells; embryonic stem cells; Endothelial Progenitor Cells; differentiated functional cells; transdifferentiated cells

Central obesity Accumulation of visceral adipose tissue related to increased incidence of metabolic disease; it is considered an important component of the metabolic syndrome

Chelation Binding or complexation of a bi- or multidentate ligand to a substrate (often a metal ion). These ligands, which are often organic compounds, are called chelants, chelators, chelating agents, or sequestering agent. The ligand forms a chelate complex with the substrate through more than one coordination site

Chemokine Small protein or peptide molecule that activates immune cells and stimulates their migration into the site of aggression

Chronic inflammation Pathological condition characterised by concurrent active inflammation resulting in tissue malfunction; the inflamed tissue is characterized by mononuclear cell infiltration (monocytes, macrophages, lymphocytes, and plasma cells), signs of angiogenesis and fibrosis

Chronobiology The study of how day/night related rhythms are influenced by living pacemakers within organisms

Circadian rhythm Biological cycle that takes about 24 hours in order to be completed

Collagen fibres Tough bundles of collagen, also called white fibres, which are the most characteristic constituent of the connective tissue, supporting tissues and providing cell structure from the outside

Complement system Complex series of blood protein whose action supports the work of antibodies

Cortisol Also known as hydrocortisone; the most important and most potent glucocorticoid in humans; produced in the outer layer of the adrenal glands

Glucocorticoids A class of steroid hormones that, in addition to playing a central role in the stress response, are involved in metabolic and anti-inflammatory responses

Cortisone Derived from cortisol; differently from cortisol, it does not bind to the mineralocorticoid receptor

C-reactive protein Acute phase protein produced by the liver and adipose tissue, that appears in plasma and is used as marker of inflammation

Crown-like structures Common formation in the adipose tissue of obese animals; several macrophages associated in a configuration that resembles a crown around dead adipocytes

Cyclooxygenases Enzymes that catalize the formation of prostanoids from arachidonic acid. Cyclooxigenases are mainly involved in inflammatory response

Cytokines Powefull chemical substances, proteins or glycoproteins, involved in cellular communication

Dendritic cell Immune cell with highly branched extensions that engulfs foreign bodies and is involved in antigen presentation to T lymphocytes

Dyslipidaemia Disruption (generally increase) in the amount of blood lipids often caused by diet and life-style habits

Eicosanoids Signalling molecules produced from oxygenation of twenty-carbon essential fatty acids that participate in inflammation or immunity, and function as messengers

Endocrine Internal or hormonal secretion; endocrine glands secrete hormones directly into the circulatory system that are involved in the regulation of metabolism, growth, development and puberty, tissue function, and also play a part in determining mood

Endoplasmic reticulum stress Situation caused by the accumulation of unfolded proteins in the endoplasmic reticulum lumen that results in cell specific responses related to repair, survival or death

Endothelial cells Unilayer of cells that line the inner surface of the vasculature. Endothelial cells are in contact with blood flow

Endothelial dysfunction Alterations on the normal biochemical processes carried out by endothelial cells, where their functions are shifted towards reduced vasodilation, a pro-inflammatory state, with prothrombic properties

Endothelial Progenitor Cells Bone marrow-derived subtype of progenitor cells with analogous properties to embryonic angioblasts, which may be recruited to the peripheral circulation and differentiate into functional mature endothelial cells, *in vitro* and *in vivo*

Estrogens A generic term for the female sex steroid hormones. In humans, estrogen is formed in the ovary, possibly the adrenal cortex, the testis and the fetoplacental unit; responsible for initiation of estrus and for the development of secondary sexual characteristics in the female

Euglycaemic clamp Also known as the euglycaemic insulin clamp, the euglycemic hyperinsulinaemic clamp, or the glucose clamp; provides steady-state measures of insulin action

Fatty acid-binding protein 4 Also called adipocyte protein-2, is a carrier for fatty acids expressed in adipocytes and macrophages

Fibroblasts Cells of connective tissue that synthesise and secrete collagen fibres, maintaining the extracellular matrix and providing a structural framework for many tissues. They occur in various shapes, such as stellate and spindle-shaped

Free Radical An atom or molecule with an unpaired electron in its outermost shell and capable of independent existence. Produced in normal cellular processes, by environmental factors (pollutants, drugs) and as by-products of the metabolism. Chemically unstable, exhibit, however, very distinct reactivities, and typically initiate autocatalytic reactions which generate more free radicals

Free Radical Scavenger Free radical inactivator that reacts with free radicals in any biological system, donating electrons or H-atoms and given rise to less reactive products derived from both, the radical and the scavenger

Ghrelin Appetite-stimulating hormone that is mainly secreted by cells lining the fundus of the human stomach

Gout A condition characterized by abnormally elevated levels of uric acid in the blood

GLUT Or SLC2 gene family are facilitated glucose transporters, being thus responsible for the downhill, passive transport of glucose across cell membranes, i.e. these transporters speed up or facilitate the equilibration of the sugar across a membrane. Prime examples include GLUT1, involved in the transport of glucose across the endothelial cells of the blood–brain barrier, and GLUT4, responsible for insulin-stimulated glucose uptake into skeletal muscle. The distribution of GLUT transporters in mammalian cells is widespread. GLUT1, GLUT3, and GLUT4 have a higher affinity for glucose, with Km values around 2 mM, whereas GLUT2 has a lower affinity for glucose (Km around 20 mM)

Growth factor Family of polypeptides or biological factors that control cellular growth, proliferation and differentiation

Hematopoietic progenitor cells Hematopoietic cells that can differentiate into mature cells, but which lack the capacity to self-renew. Hematopoietic progenitor cells are derived from hematopoietic stem cells and are intermediate to the production of mature cells

Hematopoietic stem cells Cells that have the capacity to self-renew and to differentiate into more mature cells. Give rise to all red and white blood cells and platelets; are defined by their ability to replace the bone marrow system following its obliteration (for example, by g-irradiation) and can continue to produce mature blood cells

High density lipoproteins Lipoprotein of blood plasma that is composed of a high proportion of protein with little triglyceride and cholesterol and that is associated with decreased probability of developing atherosclerosis – called also alpha-lipoprotein.

Homeobox A9 gene Sequence-specific transcription factor, which belongs to the homeobox (HOX) family of genes, plays an important role in hematopoiesis;

gene aberrant expression has been shown to be important in the development of leukaemia

Homeostasis The maintenance of a relatively constant internal environment in a living organism

Hormone Chemical substance, usually a peptide or steroid, synthesized in a gland, secreted into the body fluids and carried to target cells, which respond with an alteration in their metabolism

Hydrophilicity Refers to a physical property of a molecule that can transiently bond with water (H_2O) through hydrogen bonding. Hydrophilic molecule is one that is typically charge-polarized and capable of hydrogen bonding, enabling it to dissolve more readily in water than in oil or other hydrophobic solvents

Hydrophobicity Refers to the physical property of a molecule that is repelled from a mass of water. Hydrophobic molecules tend to be non-polar and thus prefer other neutral molecules and nonpolar solvents

Hyperglycaemia Presence of high levels of glucose in the blood plasma

Hyperinsulinaemia Excess levels of blood circulating insulin

Hyperleptinemia Elevated plasma levels of leptin

Hyperplasia Tissue growth through the increase in the number of cells

Hypertension Persistently high arterial blood pressure. Hypertension may have no known cause (essential or idiopathic hypertension) or be associated with other primary diseases (secondary hypertension). This condition is considered a risk factor for the development of heart disease, peripheral vascular disease, stroke and kidney disease

Hypertrophy Tissue growth through the increase in cell size

Hyperuricaemia Excess content of uric acid in the blood

Hypothalamus-pituitary-adrenal cortex (HPA) axis A neuroendocrine system that controls reactions to stress and regulates other physiological processes, such as digestion, immune response, mood, and energy disposal

Hypoxia Deficiency in oxygen supply in a cell, tissue or organ

Hypoxia inducible factor-α Transcription factors that respond to decrease in available oxygen in the cell

Immune cells White blood cells or leukocytes that originate from the bone marrow, including antigen presenting cells, such as dendritic cells, T and B lymphocytes, and neutrophils, among many others; defend the body against infectious disease and foreign macromolecules

Immune response Reaction of the immune system to foreign substances

Inflammation A physiological response of the organism to harmful stimuli, namely infection or tissue injury; classical signs involve redness, swelling, heat and pain

Inhibitor of nuclear factor κ B kinase Kinase that phosphorylates the inhibitor of nuclear factor κ B protein and terminates the sequestration of that nuclear transcription factor, allowing its effects on gene expression

Innate immunity Immune system function that is inborn and provides an all-purpose defence against harmful stimuli

Insulin receptor Belongs to the large class of tyrosine kinase receptors and induces a cellular response by phosphorylating proteins on their tyrosine residues. The IR is known to phosphorylate several proteins in the cytoplasm, including insulin receptor substrates (IRSs)

Insulin receptor substrate Downstream molecule in the insulin signalling pathway that is phosphorylated by insulin receptor in tyrosine residues after its activation by insulin

Insulin resistance Condition in which a higher amount of insulin is needed to produce the normal response to this hormone that often leads to the metabolic syndrome and type 2 diabetes

Insulin resistance Insulin resistance is the condition in which normal amounts of insulin are inadequate to produce a normal insulin response from fat, muscle and liver cells

Interleukin Major group of cytokines initially reported to be secreted by leucocytes with communication functions

Ischemia Relative or absolute restriction of blood supply to an organ with resultant damage or dysfunction of the tissue

Isoprenaline A synthetic beta-sympathomimetic derived from adrenaline

Jun N-terminal kinase Kinase that binds and phosphorylate c-Jun on serine residues in response to stress stimuli, such as cytokines, ultraviolet irradiation, heat shock, and osmotic shock; belong to mitogen-activated protein kinase family

Leptin 16 kDa adipose-derived protein hormone that plays a key role in regulating energy intake and energy expenditure, including appetite and metabolism; it is augmented in obese states

Leucocytes Cells of the immune system defending the body against both infectious disease and foreign materials

Lipid droplet Also called adiposome, is the lipid storage organelle in cell, especially large in adipocytes

Lipodystrophy Medical condition characterized by lack and/or altered body distribution of adipose tissue

Lipogenesis Encompasses the process of fatty acid synthesis from glucose and triglyceride synthesis, to store in lipid droplets within the cell or incorporate in VLDL particles to secrete to the blood stream

Lipolysis Breakdown of triglycerides with resulting release of free fatty acids and glycerol

Lipophilicity Refers to the ability of a chemical compound to dissolve in fats, oils, lipids, and non-polar solvents. These substances tend to dissolve in other lipophilic substances

Lipotoxicity Adverse effects resulting from the accumulation of lipids in non-adipose cells

Lipoxins Anti-inflammatory mediators derived from arachidonic acid

Low-density lipoprotein Type of lipoprotein that transports cholesterol and triglycerides from the liver to peripheral tissues. High levels of LDL are associated with an increased risk of atherosclerosis and cardiovascular disease

Lymphocyte Small white blood cell produced in lymphoid organs and essential to immune defences

Macrophage Large and versatile immune cell that phagocytises invading pathogens and other foreign bodies working both on non-specific defence or collaborating in cell-mediated immunity; are originated from blood monocytes

Mast cells Granulocyte resident in several types of tissues, whose contents in heparin and histamine contribute to the symptoms of allergy, but are also involved in wound healing

Matrix metalloproteinase Zinc-dependent endopeptidase able to degrade extracellular matrix components

Metanephrine *O*-methylated metabolite of adrenaline

Microalbuminuria Leakage of small albumin amounts into the urine

Mineralocorticoid receptors Hormone- activated transcriptional factors that regulate a wide variety of physiological processes ranging from organ development and differentiation to mood control and stress response; bind both aldosterone and cortisol with equal affinity

Mitochondrial dysfunction Within eukaryotic cells mitochondria provide most of the ATP by oxidative phosphorylation. Consequently, mitochondrial dysfunction contributes to a wide range of human pathologies, including neurodegenerative diseases, ischaemia-reperfusion injury in stroke and heart attack, diabetes and the cumulative degeneration associated with ageing. This mitochondrial dysfunction causes cell damage and death by compromising ATP production, disrupting calcium homeostasis and increasing free radical fluxes and oxidative stress

Mitogen-activated protein kinases Serine/threonine-specific protein kinases that respond to extracellular stimuli (mitogens) and regulate various cellular activities, such as gene expression, mitosis, differentiation, and cell survival/apoptosis

Monoaminoxidase Enzyme that catalyzes the degradation of naturally occurring monoamines, such as catecholamines, by oxidative deamination

Monocyte chemoattractant protein-1 Small cytokine belonging to the CC chemokine family that recruits monocytes, memory T lymphocytes and dendritic cells to sites of tissue injury and infection

Mononuclear cells Non-specific term referring to lymphocytes and plasma cells and macrophages; literally means cells without lobed nuclei (i.e. not neutrophils, eosinophils or basophils)

NAD(P)H oxidases Enzymes that catalyze oxidoreductase reactions, using oxygen as the electron acceptor molecule

Neovascularisation Formation of blood vessels *de novo*

Neuroendocrine Interaction between the nervous and endocrine system, which secretes neurotransmitter, neuromodulator or neuropeptide hormones into the peripheral circulation in response to a neural stimulus

Neutrophil White blood cell that is an abundant and important phagocyte

Nitric oxide synthase Family of enzymes that synthesize nitric oxide (NO) from L-arginine, NADPH and O_2. NOS play a crucial role in the nervous, immune and vascular systems

Non-esterified fatty acids Fatty acids released after lipolysis to the blood stream where they can be found bound to albumin

Non-phlogistic Non-inflammatory (phlogistic, from the greek *phogizo* means flame, set fire, inflammation)

Noradrenaline Also known as norepinephrine; the major neurotransmitter of the sympathetic nervous system; mainly involved in cardiovascular homeostasis

Nuclear factor κ B Transcription factor found in almost all animal cell types involved in cellular responses to stimuli such as stress, cytokines, free radicals, ultraviolet irradiation and bacterial or viral antigens; plays a key role in regulating the immune response to infection

Oxidative stress Imbalance in the dynamic equilibrium between oxidants and antioxidants in biological systems that favours the formers, potentially leading to damage. Recent updates of the concept emphasize the occurrence of compartimentalized cellular redox circuits and individual signalling, rather than global balances

Oxidized low-density lipoproteins Fraction of the LDL population that has suffered oxidative modification of its lipid and apoB protein content, increasing its atherogenic potential. The oxidation of LDL particles leads to its internalization by macrophages in the intima space of arterial wall via scavenger receptors. Upon accumulation the macrophage-loaded LDL constitute the precursors of fatty streaks. High levels of oxidized LDL appear to be also associated with three of the individual components of metabolic syndrome, such as obesity, hypertriglyceridemia, and high fasting glucose

p38 MAPK Class of mitogen-activated protein kinases which are responsive to stress stimuli, such as cytokines, ultraviolet irradiation, heat shock, and osmotic shock and are involved in cell differentiation and apoptosis

Pericytes Relatively undifferentiated mesenchymal-derived cell that supports vessels (capillaries). Pericytes can differentiate into smooth muscle cells, fibroblasts or macrophages, and play a relevant role in vessel stabilization and angiogenesis

Peroxisome proliferator-activated receptor γ Nuclear receptor protein activated by prostaglandin J2 and thiazolidinediones that functions as a transcription factor regulating the expression of genes upon heterodimerization with the retinoid X receptor and further binding to proliferator hormone response elements to regulate genes involved in adipocyte differentiation and inflammation

Peroxynitrite anion Oxidant and nitrating agent that can damage a wide array of molecules in cells, including DNA and proteins

Phagocytosis Process by which one cell engulfs another cell or a large particle

Phenylethanolamine- N- methyl transferase Enzyme found in the adrenal medulla that converts noradrenaline into adrenaline

PI3 Kinase pathway Catalyzes the production of phosphatidylinositol-3,4,5-trisphosphate, in cell survival pathways, regulation of gene expression and cell metabolism, and cytoskeletal rearrangements

Plasminogen activator inhibitor type-1 Major inhibitor of tissue (tPA) and urokinase-type (uPA) plasminogen activators, which play a relevant role in fibrinolysis. Plasminogen activator inhibitor-1 is a serine protease inhibitor (Serpin)

Polyphenols Large family of natural compounds widely distributed in plant foods; are the most abundant antioxidants in the diet

Preadipocyte Adipocyte precursor cells that resemble fibroblasts but are commited to the adipocyte lineage

Prooxidant Chemicals or substances that induce oxidative stress, either through creating reactive oxygen species or by inhibiting antioxidant systems

Protectins Family of docosahexaenoic acid-derived mediators possessing a conjugated triene structure as a distinguishing feature; involved in inflammation resolution

RAGE Member of the Immunoglobulin superfamily which binds a variety of ligands including advanced glycation end products (AGEs) and amyloid fibrils; expressed by endothelium, mononuclear phagocytes, smooth muscle and neurons. It may be involved in a range of pathological conditions including diabetes and Alzheimer's disease

Reactive oxygen species Reactive molecules derived from molecular oxygen (ROS). ROS include either free radicals (such as superoxide and hydroxyl radicals) and non-radicals (such as hydrogen peroxide). ROS play a role in cell signalling and its steady-state concentration increase during oxidative stress

Reactive nitrogen species Reactive molecules derived from nitric oxide (RNS). Formed upon reaction of nitric oxide with superoxide radical (peroxynitrite anion) or oxygen (*e.g.* nitrite radical). Are responsible for the indirect noxious effects of nitric oxide

Redox Signalling Free radical and oxidant-dependent post-translational modification of regulatory proteins that use redox chemistry and transduce an oxidant signal into a biological response

Renin Enzyme produced by the kidneys that regulates the volume of fluids in the body and blood pressure; catalyzes the convertion of angiotensinogen into angiotensin II

Renin-angiotensin system System of hormones and enzymes that plays an important role in regulating blood pressure and the body's balance of fluids and electrolytes

Resolution Complete restoration of the inflamed tissue to its normal status including carried out by anti-inflammatory mediators, neutrophil apoptosis and phagocytosis by macrophages with posterior clearance through lymphatics

Resolvins Lipid derived resolution-phase interaction products carrying bioactivity

SGLT Sodium-coupled glucose cotransporters responsible for active glucose uptake. SGLT transporters belong to the SLCA5 gene family. The most well-known member is SGLT1, which is responsible for the active transport of glucose across the brush border membrane of the small intestine. SGLT1 is a high-affinity, Na+-dependent and phloridzin-sensitive glucose co-transporter, and actively transports glucose and galactose with similar and high affinities (around 0.1-0.6 mM)

Sirtuin 1 gene Encodes for an enzyme which deacetylates proteins that contribute to cellular regulation (reaction to stressors, longevity)

Smooth muscle cells Type of nonstriated muscle cells present in the tunica media layer of arteries and veins

Somatostatin Also known as somatotropin release-inhibiting factor; a neuropeptide, widely distributed throughout the central nervous system and periphery, that acts primarily as a negative regulator of neurotransmission, cell secretion and cell proliferation

Statin Class of drugs commonly used to lower cholesterol levels by inhibiting the enzyme HMG-CoA reductase, which is the rate-limiting enzyme of the mevalonate pathway of cholesterol synthesis

Sterol regulatory element-binding protein Transcription factor involved in the activation of genes involved in lipogenesis and very low density lipoprotein excretion

Stromal Derived Factor -1 Small cytokine belonging to the chemokine family C-X-C motif ligand 12 (CXCL12), which binds to receptor CXCR4. Strongly chemotactic for lymphocytes; during embryogenesis directs the migration of hematopoietic cells from foetal liver to the bone marrow and the formation of large blood vessels; in adulthood plays an important role in angiogenesis by inducing the recruitment of Endothelial Progenitor Cells

Substance P Short-chain polypeptide that functions as a neurotransmitter especially in the pain fiber system; also involved in immune/hematopoietic system modulation

Superoxide anion Also called superoxide radical. The one-electron reduction product of molecular oxygen. Limited reactivity as compared with other more oxidizing radicals, such as hydroxyl, peroxyl and alkoxyl radicals. Usually, produced as a by-product of certain metabolic reactions but also on purpose by biological defense mechanisms

Sympatho-adrenomedullary axis Sympathetic branch of the autonomic nervous system; involved in body homeostasis, blood pressure, heart rate, energy balance and intermediary metabolism

Systemic low-grade chronic inflammation State of prolonged mild chronic inflammatory response characterised by a 2- to 3-fold increase in plasma concentrations of cytokines and acute phase proteins

Telomerase Enzyme composed of a catalytic protein component and an RNA template and that synthesizes DNA at the ends of chromosomes and confers replicative immortality to cells

Telomere Specialized nucleic acid structure found at the end of a chromosome. Associated with a characteristic DNA sequence that is replicated in a special way. A telomere counteracts the tendency of the chromosome to shorten with each round of replication

Thiazolidinediones Also called glitazones, are drugs used in the therapy of type 2 diabetes that act by binding to peroxisome proliferator-activated receptor γ and activating the transcription of specific genes involved in lipid and carbohydrate metabolism, adipocyte differentiation, angiogenesis and inflammation

Toll-like receptors Receptors that recognize microbial patterns and represent a germline encoded non-self recognition system that is involved in first line defence against pathogens

Transforming growth factor β Peptide with a role in the control of proliferation, cellular differentiation and other functions in most cells

Tumour necrosis factor α Cytokine that stimulates the acute phase reaction regulating apoptotic cell death, cellular proliferation, differentiation, inflammation, tumorigenesis, and viral replication

Type 2 diabetes mellitus Also known as non-insulin-dependent diabetes mellitus or adult-onset diabetes; is a metabolic disorder primarily characterized by insulin resistance, relative insulin deficiency, and hyperglycaemia. These alterations lead to damage and functional impairment of many organs, most importantly in the cardiovascular system

Unfolded protein response Cellular stress response related to the endoplasmic reticulum activated in response to an accumulation of unfolded or misfolded proteins in the lumen of endoplasmic reticulum; constitutes a graded series of actions related to restoration of the normal function of the cell by halting protein translation and increasing the production of chaperones, and if not enough, initiates apoptosis

Urokinase type of plasminogen activator Serine protease enzyme that triggers a proteolysis cascade by activating plasminogen into plasmin, participating hence in thrombolysis or extracellular matrix degradation

Vascular Endothelial Growth Factor Family of growth factors, VEGF or VEGF-A is the key regulator molecule of the processes of vasculogenesis and angiogenesis, during embryogenesis and adulthood. By interaction with its receptors tyrosine kinase, it mediates endothelial cell survival, proliferation, migration and differentiation during angiogenesis. Induces the mobilization and recruitment of bone marrow-derived Endothelial Progenitor Cells in vasculogenesis

Vascular repair Endogenous capacity to restore the functional integrity of a damaged endothelial monolayer, by the activation of the processes of angiogenesis and adult vasculogenesis

Vasculogenesis *De novo* vessel formation occurring during embryonic development in which angioblasts differentiate in endothelial cells that assemble into the primary capillary plexus of the embryo

Very low density lipoproteins Type of lipoprotein involved in the transport of endogenous lipids (triglycerides, phospholipids, cholesterol and cholesteryl esters), from the liver into extrahepatic tissues

Index

Kolibri 360 GmbH
Ferdinand-Jühlke-Straße 7
99095 Erfurt, Deutschland
produktsicherheit@kolibri360.de